Lecture Notes in Computer Science 6990

Commenced Publication in 1973
Founding and Former Series Editors:
Gerhard Goos, Juris Hartmanis, and Jan van Leeuwen

Editorial Board

David Hutchison
  Lancaster University, UK
Takeo Kanade
  Carnegie Mellon University, Pittsburgh, PA, USA
Josef Kittler
  University of Surrey, Guildford, UK
Jon M. Kleinberg
  Cornell University, Ithaca, NY, USA
Friedemann Mattern
  ETH Zurich, Switzerland
John C. Mitchell
  Stanford University, CA, USA
Moni Naor
  Weizmann Institute of Science, Rehovot, Israel
Oscar Nierstrasz
  University of Bern, Switzerland
C. Pandu Rangan
  Indian Institute of Technology, Madras, India
Bernhard Steffen
  TU Dortmund University, Germany
Madhu Sudan
  Microsoft Research, Cambridge, MA, USA
Demetri Terzopoulos
  University of California, Los Angeles, CA, USA
Doug Tygar
  University of California, Berkeley, CA, USA
Moshe Y. Vardi
  Rice University, Houston, TX, USA
Gerhard Weikum
  Max Planck Institute for Informatics, Saarbruecken, Germany

Abdelkader Hameurlain   Josef Küng
Roland Wagner   Christian Böhm
Johann Eder   Claudia Plant (Eds.)

Transactions on Large-Scale Data- and Knowledge-Centered Systems IV

Special Issue on Database Systems
for Biomedical Applications

Editors-in-Chief

Abdelkader Hameurlain
Paul Sabatier University, Institut de Recherche en Informatique de Toulouse (IRIT)
118, route de Narbonne, 31062 Toulouse Cedex, France
E-mail: hameur@irit.fr

Josef Küng
Roland Wagner
University of Linz, FAW
Altenbergerstraße 69, 4040 Linz, Austria
E-mail: {j.kueng,rrwagner}@faw.at

Guest Editors

Christian Böhm
Ludwig-Maximilians-Universität München
Lehr- und Forschungseinheit für Datenbanksysteme
Oettingenstraße 67, 80538 München, Germany
E-mail: boehm@dbs.ifi.lmu.de

Johann Eder
Alpen Adria Universität Klagenfurt, Institut für Informatik-Systeme
Universitätsstraße 65-67, 9020 Klagenfurt, Austria
E-mail: eder@isys.uni-ku.ac.at

Claudia Plant
Florida State University, Department of Scientific Computing
400 Dirac Science Library, Tallahassee, FL 32306-4120, USA
E-mail: cplant@fsu.edu

ISSN 0302-9743 (LNCS)   e-ISSN 1611-3349 (LNCS)
ISSN 1869-1994 (TLDKS)
ISBN 978-3-642-23739-3   e-ISBN 978-3-642-23740-9
DOI 10.1007/978-3-642-23740-9

Springer Heidelberg Dordrecht London New York
Library of Congress Control Number: 2011935603
CR Subject Classification (1998): H.2, I.2.4, H.3, H.4, J.1, H.2.8

© Springer-Verlag Berlin Heidelberg 2011
This work is subject to copyright. All rights are reserved, whether the whole or part of the material is
concerned, specifically the rights of translation, reprinting, re-use of illustrations, recitation, broadcasting,
reproduction on microfilms or in any other way, and storage in data banks. Duplication of this publication
or parts thereof is permitted only under the provisions of the German Copyright Law of September 9, 1965,
in ist current version, and permission for use must always be obtained from Springer. Violations are liable
to prosecution under the German Copyright Law.
The use of general descriptive names, registered names, trademarks, etc. in this publication does not imply,
even in the absence of a specific statement, that such names are exempt from the relevant protective laws
and regulations and therefore free for general use.

Typesetting: Camera-ready by author, data conversion by Scientific Publishing Services, Chennai, India
Printed on acid-free paper
Springer is part of Springer Science+Business Media (www.springer.com)

Preface

Life sciences, biology and medicine are highly dynamic and data intensive disciplines. During recent years, the character of basic research and good practice has dramatically changed. Advanced data acquisition technology enables unique insights into complex biological processes and diseases. Gene sequencing and protein and metabolite profiling have revolutionized biology. Modern imaging and signal recording techniques like electrophysiology, mass spectrometry and magnetic resonance imaging allow researchers to study various aspects of biomedical systems at different levels of granularity, ranging from single cells to entire organisms. A large number of databases provide public access to literature, tools and experimental data. In principle, there has never been more information available to comprehensively answer high-level biomedical research questions, like how does the human brain distinguish different objects? Or how will a certain flu epidemic evolve? However, it is usually the case that only a tiny part of the relevant information to answer such a question is accessible to the biomedical expert.

To avoid the situation of drowning in data but starving for knowledge, database technology needs to be closely integrated into biomedical research and clinical practice. This special issue of Transactions on Large-Scale Data- and Knowledge-Centered Systems consists of nine contributions that demonstrate by means of case studies database solutions for various biomedical applications including comparative genomics, neuroscience and the electronic health record. All papers underwent an extensive review process. Two of the papers are journal versions of the best papers of the ITBAM (International Conference on Information Technology in Bio- and Medical Applications) 2010 conference. The remaining papers authored by leading experts in the field have been obtained by an open call for papers.

The high-quality contributions of this special issue demonstrate that research in databases and research in biomedicine can achieve fruitful synergy. The contributions highlight some of the major challenges emerging from the biomedical applications that inspire and promote current database research:

1) Management, organization and integration of massive amounts of heterogeneous data. Phillip Rautenberg et al. present a data management system for electrophysiological neuroscience data. Based on the relational model, this database effectively organizes vast amounts of experimental and simulation data together with essential metadata and serves as an accessible platform for data and knowledge exchange among experts.

Alessio Bechini and Ricardo Giannini propose a solution for organizing the vast amount of data produced by genotyping laboratories. A document management system enriched with ontology-guided semantic tagging provides flexible data organization and can be efficiently integrated into the workflow of a typical genotyping laboratory.

Joao Zamite et al. present the MEDCollector, a system for multi-source epidemic data collection. The Web is a rich source of real time epidemic data. However, the data is available in a wide range of formats and quality. Characterized by interoperability and modularity, MEDCollector enables data collection from multiple sources. Ontologies and workflows guarantee the consistency of the integrated information.

2) Bridging the semantic gap between high-level research questions and low-level data. Heri Ramampiaro and Chen Li focus on the special challenges of information retrieval in biomedical literature databases. Traditional information retrieval techniques suffer from the highly domain-specific and ambiguous terminology in biomedical texts. To meet the special requirements of biomedical documents, the BioTracer system suitably recombines and extends existing information retrieval technology and strongly involves user feedback.

Idir A. Amarouche and colleagues propose an architecture for automatic composition of electronic health records from diverse data sources. Based on data as a service and an effective query rewriting approach, the system can handle high-level queries of medical practitioners.

Paola Bonfante and colleagues introduce the BIOBITS system enabling users to discover the evolutionary relationship between different publicly available genomes. Based on modular database architecture, BIOBITS allows the user to flexibly analyze the data over different levels of abstraction ranging from single genes to large sections of the genome.

Andreas Dander et al. present KD3, a software suite for workflow-based exploration of biomedical data. The software assists the biomedical expert in all steps of the knowledge discovery process guided by a concrete biomedical research question. As an example workflow, the authors illustrate the discovery of biomarker candidates for liver disease from breath gas analysis.

3) Privacy and efficiency. Systems biology research involves complicated heterogeneous workflows with distributed databases, online tools and local software. Also for privacy reasons, remote services and databases allow only limited customization, which causes an overhead in computation time and network traffic. Hasan Jamil presents a novel collaborative model for data integration in life sciences respecting the special privacy requirements and preserving the autonomy of remote resources.

Finally, Imen Bouraoui proposes a novel feature extraction technique for iris recognition. The human iris provides a high potential for reliable personal identification. The proposed approach is comparable in performance to the state-of-the-art in identification accuracy while being more efficient.

We thank Abdelkader Hameurlain, Josef Küng und Roland Wagner, the editors of TLDKS, for giving us the opportunity to serve as guest editors of this special issue. We further thank the reviewers for their effort and constructive suggestions. Special thanks also to Gabriela Wagner for supporting us with the organization. We hope that you enjoy the papers and perhaps find some inspiration for your interdisciplinary work in this exciting area.

June 2011
Christian Böhm
Johann Eder
Claudia Plant

Editorial Board

Hamideh Afsarmanesh	University of Amsterdam, The Netherlands
Francesco Buccafurri	Università Mediterranea di Reggio Calabria, Italy
Qiming Chen	HP-Lab, USA
Tommaso Di Noia	Politecnico di Bari, Italy
Georg Gottlob	Oxford University, UK
Anastasios Gounaris	Aristotle University of Thessaloniki, Greece
Theo Härder	Technical University of Kaiserslautern, Germany
Sanjay Kumar Madria	University of Missouri-Rolla, USA
Vladimir Marik	Technical University of Prague, Czech Republik
Dennis McLeod	University of Southern California, USA
Mukesh Mohania	IBM India, India
Tetsuya Murai	Hokkaido University, Japan
Gultekin Ozsoyoglu	Case Western Reserve University, USA
Oscar Pastor	Polytechnic University of Valencia, Spain
Torben Bach Pedersen	Aalborg University, Denmark
Günther Pernul	University of Regensburg, Germany
Colette Rolland	Université Paris1 Panthéon Sorbonne, CRI, France
Makoto Takizawa	Seikei University, Tokyo, Japan
David Taniar	Monash University, Australia
Yannis Vassiliou	National Technical University of Athens, Greece
Yu Zheng	Microsoft Research Asia, China

Reviewers

Can Altinigneli	University of Munich, Germany
Klaus Hahn	HMGU Helmholtz Center Munich, Germany
Xiao He	University of Munich, Germany
Bettina Konte	University of Munich, Germany
Grigorios Loukides	Vanderbilt University, USA
Son T. Mai	University of Munich, Germany
Rosa Meo	University of Torino, Italy
Oscar Pastor Lopez	Univ. Politecnica de Valencia, Spain
Michael Plavinski	University of Munich, Germany
Andrew Zherdin	University of Munich, Germany

Table of Contents

Database Systems for Biomedical Applications

A Database System for Electrophysiological Data 1
   Philipp L. Rautenberg, Andrey Sobolev, Andreas Herz, and Thomas Wachtler

Management of Genotyping-Related Documents by Integrated Use of Semantic Tagging 15
   Alessio Bechini and Riccardo Giannini

MEDCollector: Multisource Epidemic Data Collector 40
   João Zamite, Fabrício A.B. Silva, Francisco Couto, and Mário J. Silva

Supporting BioMedical Information Retrieval: The BioTracer Approach 73
   Heri Ramampiaro and Chen Li

Electronic Health Record Data-as-a-Services Composition Based on Query Rewriting 95
   Idir Amine Amarouche, Djamal Benslimane, Mahmoud Barhamgi, Michael Mrissa, and Zaia Alimazighi

A Modular Database Architecture Enabled to Comparative Sequence Analysis 124
   Paola Bonfante, Francesca Cordero, Stefano Ghignone, Dino Ienco, Luisa Lanfranco, Giorgio Leonardi, Rosa Meo, Stefania Montani, Luca Roversi, and Alessia Visconti

[KD^3] A Workflow-Based Application for Exploration of Biomedical Data Sets 148
   Andreas Dander, Michael Handler, Michael Netzer, Bernhard Pfeifer, Michael Seger, and Christian Baumgartner

A Secured Collaborative Model for Data Integration in Life Sciences 158
   Hasan Jamil

Flexible-ICA Algorithm for a Reliable Iris Recognition 188
   Imen Bouraoui, Salim Chitroub, and Ahmed Bouridane

Author Index 209

A Database System for Electrophysiological Data

Philipp L. Rautenberg, Andrey Sobolev, Andreas V.M. Herz,
and Thomas Wachtler

German Neuroinformatics Node, Department Biologie II,
Ludwig-Maximilians-Universität München
Grosshaderner Str. 2, 82152 Planegg-Martinsried, Germany
{philipp.rautenberg,andrey.sobolev,
andreas.herz,thomas.wachtler}@g-node.org
http://www.g-node.org/

Abstract. Scientific progress depends increasingly on collaborative efforts that involve exchange of data and re-analysis of previously recorded data. A major obstacle to fully exploit the scientific potential of experimental data is the effort it takes to access both data and metadata for application of specific analysis methods, for exchange with collaborators, or for further analysis some time after the initial study was completed. To cope with these challenges and to make data analysis, re-analysis, and data sharing efficient, data together with metadata should be managed and accessed in a unified and reproducible way, so that the researcher can concentrate on the scientific questions rather than on problems of data management. We present a data management system for electrophysiological data based on well established relational database technology and domain-specific data models, together with mechanisms to account for the heterogeneity of electrophysiological data. This approach provides interfaces to analysis tools and programming languages that are commonly used in neurophysiology. It thus will enable researchers to seamlessly integrate data access into their daily laboratory workflow and efficiently perform management and selection of data in a systematic and largely automated fashion for data sharing and analysis.

Keywords: Electrophysiology, database.

1 Introduction

The human brain is one of the most complex biological systems. It contains more than 10^{11} nerve cells and 10^{15} synaptic connections, and its functional elements extend over more then ten orders of magnitude in space, from molecular pathways in individual synapses and neurons to the entire brain. Likewise, the dynamical processes that underlie brain function span orders of magnitude in time, from submillisecond molecular and cellular processes to the acquisition of life-long memories. The capabilities of the brain are based on the function of neurons, which transmit information by generating electrical signals, so called action potentials or spikes, that stimulate other neurons to which they are connected. These networks perform highly nonlinear and recurrent processing which severely limits any purely intuitive approach to understand brain function.

1.1 Cellular and Systems Neurophysiology

Trying to understand the brain's functions requires investigation of the dynamical processes that underlie, shape, and regulate neural processing at a huge range of spatial and temporal levels. The experimental approaches to investigate neural functions reflect this variety. Non-invasive methods, such as electroencephalography or functional magnetic resonance imaging can be used to investigate brain activity at the macroscopic level (Fig. 1A), but the spatial and temporal resolutions of these methods are too coarse to measure neural signals at the levels of networks of neurons or even single neurons. To gain insight into the information processing in neurons and networks, electrophysiological techniques are indispensable (Fig. 1B). There is a large variety of experimental approaches in electrophysiology. Methods range from measuring the intracellular potential of a neuron in vitro with a single pipette electrode to the recording of both electrical potentials and spiking activity from the brain of a behaving animal with several hundreds of electrodes simultaneously. Progress in recording techniques and methodological approaches lead to ever increasing data volumes and complexity of data. In addition, the increasing complexity of the experiments lead to an increasing amount of additional information (metadata) necessary to describe and analyze the data.

1.2 Electrophysiology Data

Variety and heterogeneity of neurophysiological data pose considerable challenges for the goal of developing common databases. In electrophysiological experiments, electrical signals from neurons are recorded as time-series data such

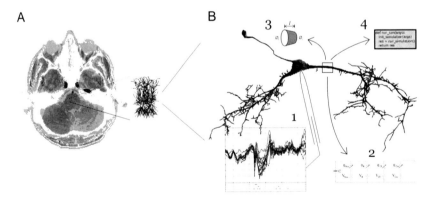

Fig. 1. *Illustration of neuroscientific data.* **A**: Collage of a horizonal section of a human brain and a network of neurons. **B**: Single neuron and sketch of a patch-clamp experiment recording currents accross the membrane to illustrate different aspects of neuroscientific data. Electrical responeses and properties like voltage traces or spike trains (1) can be recorded within living tissue. The membrane properties responsible for electrical potential of a neuron can be discribed by electric circuits (2), geometrical properties (3), mathematical equations, and by implemented software (4).

as time-varying voltage signals. In spiking neurons, information is encoded in the sequence of action potentials fired, and therefore often the times of these events are extracted from the recorded signal, yielding a representation of the activity of the neuron as spike trains. Apart from these common basic types of data, experiments can vary in virtually all experimental aspects, such as the number and types of electrodes, the species and recording site, the preparation, the stimulation paradigm, and of course all the detail parameters of the recording and stimulation equipment. To meaningfully analyze a given data set, much of this information has to be known, and to ensure reproducibility, all of it should be recorded as metadata. Organizing metadata in itself is a challenge and exchanging metadata requires standardized mechanisms such as a common format (`www.g-node.org/odml`) [8]. Regardless of how metadata are organized in detail, data management can be efficient only if the metadata and the data are available together in a unified way, such that the information provided by the metadata can be taken into account when accessing the data. Providing metadata is therefore a crucial requirement for databases of experimental data.

1.3 Data Repositories for Electrophysiology Data

Compared to other disciplines in the life sciences, such as Genomics and Proteomics [6, 14], the neurosciences lag behind regarding the use of databases for the organization and exchange of data. Only recently attempts have started to integrate neuroscience databases [1, 9], and here the focus is mostly on anatomical data. There are only few data repositories providing access to physiological data. One of the first attempts to build a database of neurophysiological data was the `neurodatabase.org` project [5], which was established 2004, funded by the NIMH Human Brain Project and created by the Laboratory of Neuroinformatics at Weill Medical College of Cornell University. In this project, an elaborate data model and format as well as a query protocol for the exchange of neurophysiological data was developed and the data are provided in this standard. The data sets available are typically data from publications and are available with an extensive amount of metadata.

In 2008, the NSF initiated the funding program Collaborative Research in Computational Neuroscience (CRCNS), which included a project to set up a repository of neurophysiological data [15]. Several laboratories received funds to document sample datasets that were made available at the crcns.org web server. Format and documentation are different for each dataset, and for most datasets stimulation data are available.

Data repositories like neurodatabase.org or crcns.org host data that have been uploaded specifically for the purpose of making the data available to the public. Typically, these data are from studies that have been published and they may be used for further investigations after they have served their primary purpose. Other projects are pursuing a different approach and provide data sharing infrastructure also for data that are not necessarily publically available. The CARMEN project (`www.carmen.org.uk`), funded by the British Engineering and Physical Sciences Research Council, provides a platform for data analysis and data exchange where the owner of the data can keep the data private, or can make the data available to selected users or the public. The platform is intended as a virtual neurophysiology laboratory that provides not only infrastructure for data exchange but also services for data analysis. Data files can be organized in a folder hierarchy and can be shared on a file-by-file basis. During file upload, the user has the option to enter metadata describing the experiment in which the data were recorded. This is done via web forms that provide fields corresponding to the minimal metadata that were proposed by the Carmen consortium [7].

The German Neuroinformatics Node is a project funded by the German Ministry of Education and Research (BMBF) that also has the goal to establish a platform for the sharing of neurophysiological data [10]. The strategy pursued is to not only set up a repository of data files, but to provide a framework that scientists can use to manage, access, and work with their data within their local workflow – in a manner that allows for data sharing with the lab, collaborators, or the public without further obstacles. A key requirement is the ability to store and access both the recorded data and the metadata (Fig. 2A) together so that all information necessary for data analysis, re-analysis, and sharing is available in a unified way. This integration of data and metadata has two benefits (Fig. 2B). First, data handling in the laboratory from recording to analysis becomes more efficient and reproducible. Second, data sharing requires no further effort because all the information is already available with the data.

1.4 Contributions

Here, we present a data management system for electrophysiological data that is based on relational database technology and domain-specific data models. We introduce mechanisms for data management that account for the heterogeneity of electrophysiological data. Furthermore, our approach provides interfaces to analysis tools and programming languages that are commonly used in neurophysiology. It thus will enable researchers to seamlessly integrate data access into their daily laboratory workflow and efficiently perform management and selection of data in a systematic and largely automatized fashion for data sharing and analysis.

In the following, we first introduce the concept of using relational database technology for data management to integrate various stages of scientific work within the electrophysiological domain. Then we address technical and implementational aspects to explain the approach in further detail and provide an example and conclusions.

2 Concepts

To build a system that can be integrated with the scientists workflow without barriers, different tools and aspects of scientific data analysis have to be taken into account: the programming language and the persistent data storage as general tool and infrastructure, but also objects and working mechanisms that help e. g. to handle analysis workflows with according results, to access meta-data, to arrange data for simulations, or to exchange data and functionality in order to collaborate with other scientists. The single datum can be compared to a juggling ball amongst many that is thrown, caught, and processed around the entire scientific workbench. In order to allow the tools to handle data in the appropriate way, the data have to be structured appropriately. Therefore, we rely on the relational model [3]: Many relational database management systems (RDBMSs) can ensure referential integrity and data consistency, and additionally the relationships are stored within the database and not as additional annotation or hidden within individual code. Furthermore, with SQL, there exists an established standard of querying relational databases and most significant programming languages provide APIs to access RDBMSs. Alternatives like NoSQL databases or object-oriented databases are more flexible or closer to programming. However, those approaches shift aspects of referential integrity or data consistency towards the application layer, whereas RDBMS inherently provides for data integrity and consistency.

The Neuroscientist within the Loop. Scientists treat their experimental data and results usually very confidential and are reluctant to share especially before the publication of results. However, at a late point in the study it often turns out to be difficult to migrate or share the data of a project, as recorded data and corresponding metadata are distributed across lab notebooks, raw data files, variables and functions within the source code, and other sources. We propose to avoid this situation and ensure data integrity by working with a structured persistent data storage from the beginning. Therefore, the design of our concept accounts for the 'scientific workbench' and extends it with tools that facilitate the use of RDBMSs.

Integrating Data Access and Analysis Code. For the analysis of (experimental) data, it is important that scientists are able to re-use code that has already been written previously. Therefore, we use an established standard as interface to a database: RDBS account for that and there are established APIs for several languages available e.g. for Matlab, Python, C++, and R which are core languages used in the area of neuroscience.

Procedures and Functions. Many procedures and functions are used to analyze data. Basic analyses like statistical descriptors or tests are used widely. Moving them into the database facilitates their application to data, their distribution to colleagues, and the automatization of entire analysis workflows.

Data Models. Specific data structures correspond to specific areas within neuroscience (see Fig. 3, respresenting data of electrophysiological experiments). The specification of such data structures facilitates the deduction of relational schemata, or source code that represents corresponding objects. We call the specific data structure together with its canonical deduction rules data model (DM). It consists of a set of objects and their relation with each other (see definition 'DM', later).

A specific data model defines a set of core objects representing raw data that are commonly used within a specific area of neuroscience. For example, within

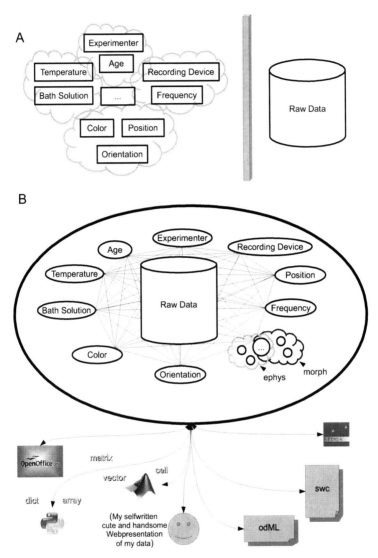

Fig. 2. *Raw data, metadata, and its access.* **A**: Raw data comes together with meta data. For electrophysiological raw data, useful metadata gives information e.g. about experimental paradigms, experimental protocols, hardware devices, laboratories, and scientists. **B**: In order to be able to use raw data in an appropriate way, a database integrates and relates data with according metadata. Here, specific fields are identified to establish templates for corresponding schemata. These schemata can overlap, as illustrated for the two examples 'ephys' (schema to respresent data of electrical properties of neurons) and 'morph' (schema to represent morphological properties of a neuron). The database provides a standard interface which can be used by software and programming languages, that are commonly used by scientists, like Matlab, Python, or OpenOffice to import, export, analyze, visualize, and publish data.

Table 1. The data model 'neo_simple' has the two object RecordingPoint and Analogsignal. One Recordingpoint can have many AnalogSignals.

object-name	attribute-name	attribute-type	constraint
RecordingPoint	recordingpoint_key	integer	primary key
AnalogSignal	analogsignal_key	integer	primary key
AnalogSignal	recordingpoint_key	integer	foreign key

Table 2. This example of a configuration conf$_{neo\_simple}$. In this example 'unit' is a data-attribute and, therefore, has to be unique: it is either a data-attribute of 'RecordingPoint' or a data-attribute of 'AnalogSignal' depending on the scenario the data describes (e.g. if the recording electrode determines unit (e.g. 'mV') 'unit(datum)' should be part of 'RecordingPoint').

object-name	attribute-name	attribute-type
RecordingPoint	scaling_factor	float
RecordingPoint	description	string
AnalogSignal	frequency	float
AnalogSignal	sampling_points	integer
AnalogSignal	unit	string
AnalogSignal	signal	vector(float)

the area of electrophysiology, this set contains recorded voltage signals, spike trains, and electrodes. For each set of raw data, there will be information that describes structure of the datasets in general and information that is specific to the individual data. To add lab specific information about objects of a data model, scientists can assign 'data attributes' to these objects. As these data attributes will differ from lab to lab, we call them a 'configuration' of a data model 'xyz' (e.g. conf$_{xyz}$).

Some attributes specify the structure via relationship constraints ('relationship attribute') while others describe data ('data attribute'). Here we provide a formal definition of a data model ($DM \in \mathcal{DM}$) and the attribute-assignment ('configuration': conf$_{DM} \in \mathcal{C}_{DM}$) of a data model (see Table 1, 2), with:

$$O := \{\text{object-names: string}\} \qquad (1)$$
$$T := \{\text{types}\} \qquad (2)$$
$$C := \{\text{relationship constraints}\} \qquad (3)$$
$$A := \{\{\text{attribute-names: string}\} \times T\} \qquad (4)$$
$$\mathcal{DM} := \mathcal{P}(O \times A \times C) \qquad (5)$$
$$\mathcal{C}_{DM} := \{C \in \mathcal{P}(O \times A), DM \in \mathcal{DM} \mid \qquad (6)$$
$$\mid \forall (o, a), (o', a') \in C : a = a' \Rightarrow \qquad (7)$$
$$\Rightarrow o = o' \land \exists (o'', a'', c'') \in DM : o'' = o\} \qquad (8)$$

Metadata. In addition to *structural metadata* (provided by the data model) and *descriptive metadata* (content of data attributes about specific instances), we distinguish between *internal and external metadata*: internal metadata about an instance originate from the same data model as the instance itself. External metadata about an instance are metadata that originates from an instance of a distinct data model.

To account for the diversity of metadata required to describe different datasets, further data models are needed to describe a set of objects representing external metadata (e.g. information about experiments, like 'scientist', 'lab', and 'experimental paradigm'). In order to keep the data models independent, a 'metadata' instance is linked to a specific object by specifying a 'gluing'-relationship (see below).

Workflow. Accessing, analyzing, and storing data are essential actions in daily scientific work. Relationships between these actions, the applied parameters, and the resulting outcome can also be stored as relations and tuples within the relational model. Relations help to manage data and to trace actions of the workflow. Parts of data management is even be done automatically by the RDBS (depending on the schema), such as keeping data consistent. On the other hand, data integrates into an individual development environment with already existing code and scripts by using object relational mappings. Scientists stay flexible and develop their own structures/schemata, using at the same time well-proven systems that keep referential integrity and data consistiency.

3 Technical and Implementational Details

We implemented our solution using PostgreSQL 9.0 as RDBMS. Among the many different technical approaches for storing, retrieving, and managing of data, PostgreSQL is a powerful, open source object-relational database system. Its SQL implementation strongly conforms to the ANSI-SQL:2008 standard and, therefore, is widely compatible with other database systems. With version 9, PostgreSQL supports hot standby and replication which increases possibilities with respect to backup and performance. Further improvements important for neuroscientific applications have been introduced, such as the removal of limits of key and value sizes for 'hstore', which represents sets of (key, value) pairs within a single data field. This is useful in various scenarios like storing semi-structured metadata that are rarely examined. Furthermore, the support for 'arrays' as parameters and return values to other procedural languages has been enhanced. This is especially important as many neuroscientific analysis tools work with vectors, which can be mapped to PostgreSQL-arrays.

PostgreSQL supports Python (PL/Python) and R (PL/R). These two programming languages bring the database closer to the neuroscientist: Python has become a de facto standard in several areas for which most neuroscientific research simulators like NEURON, or NEST provide interfaces [2]. It provides also powerful packages for data analysis (e.g.: numpy, scipy) and plotting (e.g.

matplotlib). R is used widely for statistical computing and graphics [13]. Together with these and other programming languages the RDBMS can offer a wide range of functionality that the scientist needs and to fine-tune the interface to the individual workspace.

From the software development point of view it is also very interesting that PostgreSQL supports multiple schemata which separate database objects like tables, views, sequences, and functions into different name-spaces. This enables scientists to divide the database into logically distinct parts while using constant connection to the server. Furthermore, schemata add another organization level to handle rules and permission, which facilitates the management of many users accessing one database for collaboration, or global data access. Practically, each user has an individual schema for each project with an administrator-role for this schema.

In order to keep the neuroscientist within the loop and allow for smooth code-cooperation between scientists, we suggest some conventions for schema structures and data models. This also lowers entry barriers for new features, which will appear immediately familiar if they adhere to the same conventions. Those conventions are being developed and extended according to the requirements of the community. For example, all names of DB-objects that belong to a specific data model start with the name of the model followed by '_' (e. g.: 'neo_', see example, later), or primary keys consists only of a single column of integers.

3.1 Data Model with Data and Internal Metadata

To define a data model of core objects within the area of electrophysiological raw data, we adopt the approach developed in the `neo` project (`neo`: Neural Ensemble Object, see `http://neuralensemble.org/trac/neo`). This project provides common class names and concepts for dealing with electrophysiological (experimental and/or simulated) data. One aim is to provide a basis for interoperability and integration of projects like OpenElectrophy [4], NeuroTools (`http://neuralensemble.org/trac/NeuroTools`), G-Node [10], and other projects with similar goals [12].

The central data objects in the area of electrophysiology are the time series, such as voltage traces recorded from neurons. `neo` represents time series by the AnalogSignal object. But the `AnalogSignal` is useless without its corresponding metadata. `neo` provides this internal metadata in two canonical ways: each instance of an object (like `AnalogSignal`) is embedded into a hierarchical structure which provides the necessary implicit metadata in order to relate this instance with other instances correctly. Those other related instances (e.g. an instance of `RecordingPoint`) provide explicit metadata together with their data attributes (e.g. 'scaling_factor', or 'polarity', 'recording_unit') . The design of this hierarchical structure reflects a typical setup of an experimental paradigm: one or more electrodes (`RecordingPoint`) measure simultaneously for a specific period of time (`Segment`) the activity (`SpikeTrain`, `AnalogSignal`) of neurons (`SingleUnit`). To identify a specific `AnalogSignal`, the corresponding `RecordingPoint` and `Segment` could serve as space-time coordinates, thus using metadata to select raw data.

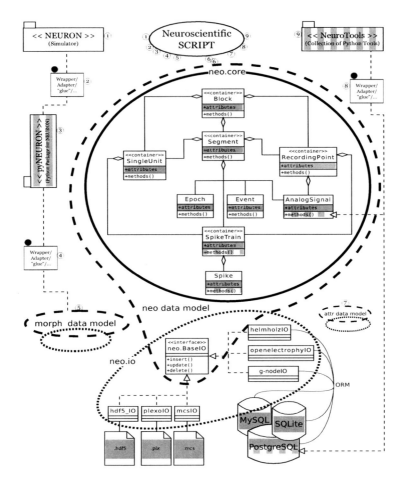

Fig. 3. *Integrating specific data models into the workflow of neuroscientists.* Neuroscientists use different open source tools. Here we illustrate the data model **neo** and how different developments within the neuroscientific community can be supported and integrated into the neuroscientific workflow. **neo.core** describes a non-cyclic hierarchical relation between classes. The hierarchy reflects the structure of a typical experimental paradigm: In an experimental session (**Block**) during a trial (**Segment**), one or more electrodes (**RecordingPoint**) measure simultaneously the voltage traces of neurons (**AnalogSignals**). Analyzing these **AnalogSignals** of **RecordingPoints**, **Spikes** are detected and assigned to **SingleUnits** (corresponding to a neuron, causing the **SpikeTrain**). **Events** reflect a point in time (**Epoch**: time period) where e.g. an stimulus was presented. We can map data (grey) to different file formats but also to relational databases. Neuroscientists analyze data using their custom software (*neuroscientific SCRIPTS*). To keep the individual workbench as independent as possible, specific analysis tools can be integrated into the database. For example, in a PostgreSQL-instance entire simulators like pyNEURON [11] or Analysis packages like 'NeuroTools' (both based on Python: grey stripes) can be used because PostgreSQL supports Python as procedural language.

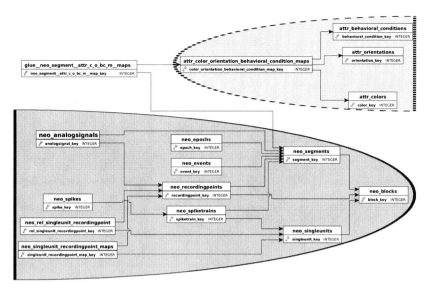

Fig. 4. *'Gluing' different data models on the backend side.* The schema at the left bottom reflects the data model **neo** whereas the schema on the right top reflects the data model 'attr' which stores data about controlled variables of the experimental protocol. Both models stay unchanged to keep software dependencies valid. They are related by an additional relation which is 'gluing' them together. With this extension at the backend and an according extention at the script level, scientists can analyze **neo**-data taking corresponding attr-data into account without changing the core structure of the individual data models.

This illustrates how data model can reflect data acquisition. For data analysis, the neuroscientific community developed packages (e.g. 'NeuroTools') that provide tools to analyze this data. In Fig. 3 we illustrate the structure for integrating these packages, **neo**, and the data storage.

3.2 Extending the Model

Our approach is highly modular by keeping data models independent and providing a way to relate different data models with each other. This supports a decentralized development that is at the same time easily extensible for individual requirements. Different data models are combined by 'glues' which is not modifying the structure of the data models but supplements structure about the relationship between the data. Fig. 4 exemplifies this concept on the level of a database schema: the data model 'attr' stores data about the attributes of an experimental paradigms and is glued to the **neo** data model.

3.3 Application and Examples

Through the SQL interface the data are accessible from various applications. This enables the researcher to integrate data storage and data management with the analysis tools used in the laboratory (Fig. 5). Since not only the recorded data

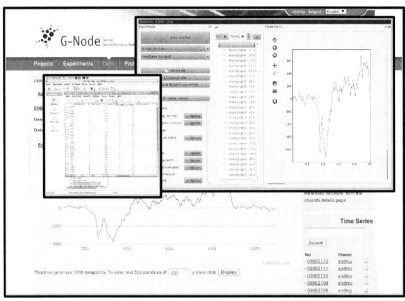
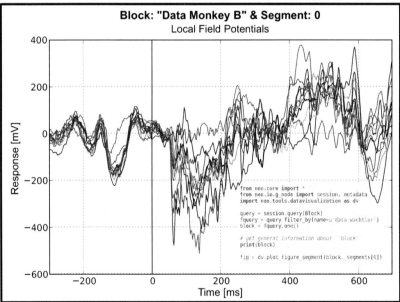

Fig. 5. *Accessing the data from different applications.* **Top:** Data visualization and analysis tools can use the database interface to access both data and metadata. The figure shows different applications (the web-based G-Node data platform, OpenElectrophy, and OpenOffice) accessing and/or plotting the same data, a recorded voltage signal, from the same database. **Bottom:** Using a specified data model establishes the possibility to write adjusted modules that use structural and descriptive metadata in order to collect all information needed e.g. for plotting the data.

but also all the metadata are accessible in this unified way, all information about the data is available. This provides high efficiency for the entire data analysis workflow, from first inspection of the recorded data, to adequate selection of data according to experimental parameters, and even to the step of writing the publication, for which the necessary information about the experiment is readily available.

4 Conclusion and Outlook

Neurosciences are faced with increasing amounts and complexity of experimental data. To cope with these challenges and to make data analysis, re-analysis, and data sharing efficient, data together with metadata should be managed and accessed in a unified and reproducible way, so that the researcher can concentrate on the scientific questions rather than on problems of data management. We have presented a solution to this challenge that builds on database technology to provide access to all information necessary to meaningfully analyze electrophysiological data already at the local neuroscientific workbench.

We built a system that can be integrated to the scientists workflow without barriers and that keeps the scientist within the loop. As a key technology we use a RDBMS, more precisely a PostgreSQL server. It ensures referential integrity and data consistency, and additionally the relationships are stored within the database. It allows at the same time to use function and triggers written in Python or R, which are core languages within the neuroscientific community.

We illustrated the application of our approach with the data model `neo` which was developed by the community (http://www.neuralensemble.org/) for dealing with neurophysiological data. In addition, we demonstrated how neuroscientists can extend it in a modular way.

This approach will enable researchers to seamlessly integrate data access into the laboratory workflow, exchange data with collaborators, and efficiently perform management and selection of data in a systematic and largely automated fashion for data sharing and analysis.

Acknowledgments. We thank Holger Blasum, Felix Franke, Christian Garbers, Nina Hubig, Oliver Janitza, Christian Kellner, Philipp Meier, Roman Moucek, Raphael Ritz, Michael Schmuker, Adrian Stoewer, Michael Stransky and Alvaro Tejero Cantero for discussions and helpful criticism, the Neural Ensemble group for discussions on the neo objects, and the open source community for providing great software. Supported by BMBF Grant 01GQ0801.

References

1. Amari, S.-I., Beltrame, F., Bjaalie, J.G., Dalkara, T., De Schutter, E., Egan, G.F., Goddard, N.H., Gonzalez, C., Grillner, S., Herz, A., Hoffmann, K.-P., Jaaskelainen, I., Koslow, S.H., Lee, S.-Y., Matthiessen, L., Miller, P.L., Da Silva, F.M., Novak, M., Ravindranath, V., Ritz, R., Ruotsalainen, U., Sebestra, V., Subramaniam, S., Tang, Y., Toga, A.W., Usui, S., Van Pelt, J., Verschure, P., Willshaw, D., Wrobel, A.: Neuroinformatics: the integration of shared databases and tools towards integrative neuroscience. Journal of Integrative Neuroscience 1(2), 117–128 (2002)

2. Cannon, R.C., Gewaltig, M.-O., Gleeson, P., Bhalla, U.S., Cornelis, H., Hines, M.L., Howell, F.W., Muller, E., Stiles, J.R., Wils, S., De Schutter, E.: Interoperability of Neuroscience Modeling Software: Current Status and Future Directions. Neuroinformatics 5(2), 127–138 (2007)
3. Codd, E.F.: A relational model of data for large shared data banks. 1970. M.D. Computing: Computers in Medical Practice 15(3), 162–166 (1970)
4. Garcia, S., Fourcaud-Trocmé, N.: OpenElectrophy: An Electrophysiological Data- and Analysis-Sharing Framework. Frontiers in neuroinformatics 3(May) 14 (2009)
5. Gardner, D., Knuth, K.H., Abato, M., Erde, S.M., White, T., DeBellis, R., Gardner, E.P.: Common data model for neuroscience data and data model exchange. J. Am. Med. Inform. Assoc. 8(1), 17–33 (2001)
6. Gelbart, W.M., Crosby, M., Matthews, B., Rindone, W.P., Chillemi, J., Russo Twombly, S., Emmert, D., Ashburner, M., Drysdale, R.A., Whitfield, E., Millburn, G.H., de Grey, A., Kaufman, T., Matthews, K., Gilbert, D., Strelets, V., Tolstoshev, C.: Flybase: a drosophila database. the flybase consortium. Nucleic Acids Res. 25(1), 63–66 (1997)
7. Gibson, F., Austin, J., Ingram, C., Fletcher, M., Jackson, T., Jessop, M., Knowles, A., Liang, B., Lord, G., Pitsilis, P., Periorellis, P., Simonotto, J., Watson, P., Smith, L.: The carmen virtual laboratory: Web-based paradigms for collaboration in neurophysiology. In: 6th International Meeting on Substrate-Integrated Microelectrodes 2008 (2008)
8. Grewe, J., Wachtler, T., Benda, J.: odML format and terminologies for automated handling of (meta)data. In: Front. Neurosci. Conference Abstract: Neuroinformatics 2010 (2010)
9. Gupta, A., Bug, W., Marenco, L., Qian, X., Condit, C., Rangarajan, A., Müller, H.M., Miller, P.L., Sanders, B., Grethe, J.S., Astakhov, V., Shepherd, G., Sternberg, P.W., Martone, M.E.: Federated access to heterogeneous information resources in the neuroscience information framework (NIF). Neuroinformatics 6(3), 205–217 (2008)
10. Herz, A.V.M., Meier, R., Nawrot, M.P., Schiegel, W., Zito, T.: G-Node: an integrated tool-sharing platform to support cellular and systems neurophysiology in the age of global neuroinformatics. Neural Netw. 21(8), 1070–1075 (2008)
11. Hines, M.L., Davison, A.P., Muller, E.: NEURON and Python. Frontiers in neuroinformatics 3(January), 1 (2009)
12. Ljungquist, B., Petersson, P., Schouenborg, J., Johansson, A.J., Garwicz, M.: A novel framework for storage, analysis and integration through mediation of large-scale electrophysiological data. In: 5th International IEEE/EMBS Conference on Neural Engineering (2011)
13. Paradis, E., Claude, J., Strimmer, K.: APE: Analyses of Phylogenetics and Evolution in R language. Bioinformatics 20(2), 289–290 (2004)
14. Stoesser, G., Sterk, P., Tuli, M.A., Stoehr, P.J., Cameron, G.N.: The embl nucleotide sequence database. Nucleic Acids Res. 25(1), 7–14 (1997)
15. Teeters, J.L., Harris, K.D., Jarrod Millman, K., Olshausen, B.A., Sommer, F.T.: Data sharing for computational neuroscience. Neuroinformatics 6(1), 47–55 (2008)

Management of Genotyping-Related Documents by Integrated Use of Semantic Tagging

Alessio Bechini[1] and Riccardo Giannini[2]

[1] Univ. of Pisa, Dept. of Information Engineering, largo Lazzarino 56126 Pisa, Italy
a.bechini@ing.unipi.it
[2] Univ. of Pisa, Dept. of Surgery, via Paradisa, 2 56124 Pisa, Italy
r.giannini@med.unipi.it

Abstract. A widespread need is present in molecular biology laboratories for software systems to support the internal management of data and documents. A typical case is represented by genotyping procedures, which produce a large amount of documents whose content may represent a potentially important knowledge base. The exploitation of such information requires a proper classification of the elements in the knowledge base, and this can be effectively achieved using concepts and tools from research on the Semantic Web. In particular, genotyping-related documents can be handled through a DMS (Document Management System) that is also able to deal with semantic metadata, e.g. in the form of tags. The use of semantic tagging at this operating level is currently hampered by the lack of proper tools. In this paper, based on experience from a practical case, we present an integrated approach to manage relevant genotyping documents and to deal with their semantic tagging. A preliminary study on the test procedures workflow is crucial to understand the document production processes. The employed semantic annotation makes use of terms taken from domain ontologies in the biomedical field. The annotation tool must be seamlessly integrated in the supporting DMS; the tool flexibility and usability guarantee a low overhead for the annotation process, paving the way for a widespread adoption of semantic tagging for genotyping-related documents.

Keywords: Laboratory Information Management Systems, document management, semantic tagging, biomedical ontologies, biomolecular test workflow.

1 Introduction

Genotyping procedures used in a biomolecular laboratory produce a large amount of data and documents, and it is difficult to manage such information in a disciplined way [28]. This task can be performed by resorting to Laboratory Information Management Systems (LIMS) [21,28,27,17], which mainly deal with keeping track of samples and with capturing data at different steps of the overall procedure. Conversely, document management so far has not deserved specific attention in this context. The scenario is further complicated by the fact that heterogeneous file formats are employed (often depending on the available instrumentation) and by the possible lab-specific workflow followed to obtain the final results.

In addition to the basic requirements that data and documents have to be properly collected and that tracing issues have to be guaranteed, it is important to design and implement a data/document repository whose elements could be effectively accessed and searched. It has been shown in recent literature that an organized structuring of genotype-like data is crucial to foster new findings [18,33], especially in association study analysis; this observation pushes to investigate on better solutions for information classification and retrieval. In the field of research literature, the central ideas of the Semantic Web [10] have found significant application to the management of biomedical publications (and research results in general). The biomedical research community has increasingly experienced the need of a shared precise terminology, and not surprisingly the influence of Semantic Web principles as contributed to an amazing growth of standardized controlled vocabularies and ontologies [11]. The possibility to formally codify established knowledge into ontologies may also represent a means and an opportunity to relate findings in a publication to a precise network of grounded concepts. This kind of relation is obtained through *semantic annotation*. A semantically annotated corpus of publications and, in general, research results, is a massive knowledge base whose full exploitation is likely still far to come [32]. An even more substantial advantage could come from the semantic classification of heterogeneous documents from research, or even from ordinary diagnostic tests. In fact, it is reasonable to maintain a semantically searchable knowledge base at the laboratory level, e.g. using a "semantically-enabled" LIMS, at least for the documents produced during test executions [9]. In a wider perspective, a common metadata framework for such a kind of knowledge bases can be definitely considered as an enabling technology for their exploitation in a coordinated and integrated way. Standardization of access procedures to Document Management Systems (DMS) has recently deserved particular attention from the involved industrial players, e.g. leading to the definition of the CMIS directives [13]; such efforts are paving the way to the actual federation (at least under the search perspective) of locally developed document bases. Following this approach in the biomedical domain, researchers and clinicians might be given a new extraordinary source of information.

Semantic search can be carried out only if the documents in the knowledge base have been properly annotated with unambiguous, standard terms [15]. Standardization efforts in this direction lead to the development of controlled vocabularies as the well-known MeSH [2] in the biomedical field. This kind of document annotation (or tagging) turns to be useful if it is performed in a very accurate way, so it requires both time and expertise from the tagging operators. Semantic tagging at this operating level has been deemed unpractical because of the difficulties in selecting the correct terms within complex ontologies [11], and also because of the lack of proper tools. Genotyping tests can be considered a paradigmatic case in this setting, because they require different instruments through the progression of the test procedures, and different types of documents are produced along this path. A general way to deal with such documents may rely on a Document Management System (DMS), which provides specific functionalities to deal with issues around archiving, classification and search. As we are interested in wide-range exploitation of our documents, we suppose that our DMS would be embedded in an Enterprise Content Management system (ECM), which give us multiple additional ways to make use of (and possibly share) our documents. The archiving in the ECM can be regarded as the proper point to introduce annotations. In this work, we propose to use semantic tagging on documents produced throughout ordinary genotyping test procedures; this activity must be integrated in the tools employed to manage such documents. Such an approach can be successful only if the tagging procedure would be practical to the end-user. Thus the supporting software tools have to be handy, and they are asked to be as less intrusive as possible with respect to the ordinary course of the archiving activities. In other words, the supporting tools must be able to smoothly introduce the semantic tagging features in the used ECM. The system described hereafter has been developed to meet this kind of requirements.

In this paper, an approach to manage and semantically annotate documents throughout genotyping procedures is presented. After a review of related works, Section 3 is devoted to a formal analysis of the process workflow. Section 4 discusses issues in document annotation, considering both the semantic viewpoint and the particular characteristics of the biomedical domain. The description of the proposed approach and the structure of the supporting software framework are given in Section 5, and details on the actual employment in a real-world, typical case can be found in Section 6. Conclusions are drawn in Section 7.

2 Related Works

The daily activity of a biomolecular laboratory strongly needs software systems to capture and handle the generated data and documents, as well as to track samples [21]. Moreover, sometimes also the specific steps in the overall genotyping activities should be formally described [23]. LIMS have been proposed as a solution for these problems, and they have been built up out of simple tools to make them usable by as many users as possible (even by those with minimal informatic skills). A natural, basic instrument to support data keeping is a DBMS; its adoption in a very simple way has been proposed in early works [27], and later it has been further developed in more organic structures [21]. In other cases, a collection of supporting programs (or macros) has been created [28]. Anyway, there exists a common agreement in stating that a proper software suite is an invaluable help in supporting quality control. Recently, efforts have been spent in developing comprehensive frameworks to be adapted to each particular laboratory setting [20].

The main focus in LIMS has been the support to the laboratory procedures, and minimal emphasis has been put to the management of potential value of the collected documents. Some hints on how to tackle this last issue may come from research on the Semantic Web.

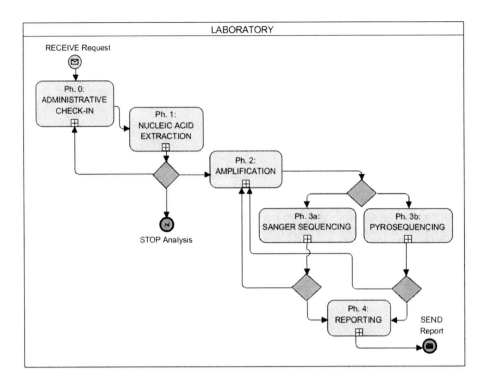

Fig. 1. A BPMN description of the overall procedure addressed in the work. Each test session is usually started upon a pool of requests, and it operates on the corresponding biological samples. The BPMN gateways (diamond-shaped symbols) refer to quality checks on outcomes.

Founding ideas of Semantic Web have received widespread interest in the biomedical research community. Approaches to data and document management in this environment have been deeply influenced by the vision of semantic searches [10,15], and also bio-ontologies have been recognized as a key resource to organize and standardize knowledge [38,11]. Following this mainstream in biomedical data management, comprehensive semantic models have been proposed primarily to tackle interoperability issues [14]. So far, these efforts have produced conceptual models suited to the design of new information management systems, rather than fitting the structure of existing software tools. A typical example of the application of ontologies in a web-based tool for biological investigation is an early work on the "RiboWeb" system [3], dedicated to the study of the ribosome: it used an ontological approach to formally represent different features of the experiments that can be performed online, using computational resources over available datasets. The collaborative aspects in this kind of tools immediately come into evidence. It is worth underlying that researchers in molecular biology, and particularly those involved in genotyping technologies, have soon recognized the need to rely on a supporting database, properly structured to ease the automation of data feeding [26,21]. Anyway, methods for data storage have been considered so far as mere side activities respect to the investigation goals [26], and a cross-field vision is often missing. To the best of our knowledge, we are not aware of Laboratory Information Management Systems (LIMS) that leverage ontology terms for tagging documents produced along the development of bio-molecular tests, despite the fact that a compelling need for a precise structuring of information from biomedical experiments has been clearly pinpointed and addressed [24]. Anyway, approaches similar to the one shown in this paper have been used in very different domains, for example ebXML repositories for enterprise documentation [8]. In this last case, the system has shown to be both user-friendly and effective for document archival and retrieval, although the available domain ontologies were really less abundant.

Regarding the development and standardization of both foundation and domain ontologies to be used in biomedical research, a lot of work has been done and organizations exist to take care of definition and maintenance of bio-ontologies (in particular, OBO [35] is worth being recalled).

3 Analyzing Genotyping Tests Workflow

The analysis of the actual workflow carried out to complete a genotyping test is an integral part of the methodology followed in this work. Such analysis is important because, although the overall procedures are clearly defined, their application can be slightly different in each laboratory. In particular, the co-existance of different types of test procedures may lead to accommodate the workflow in order to speed up the average test completion latency (or throughput). The workflow analysis is aimed at identifying the produced documents and data that are required/worth being archived and possibly tagged with standard ontology terms. We have chosen BPMN [30] as the conceptual tool for workflow modeling. It is a formal, graphical, and self-evident notation suitable to model general business and production processes [40], and several software applications are currently available to support BPMN-based graphical modeling and simulation.

The present work takes a specific real-world case as a reference, and most of the following models are based on it: our target lab mainly deals with samples of neoplastic tissues, and it is specialized in detecting mutations that affect the choice of the pharmacological treatment. Although the discussion may benefit from the reference to a concrete setting, all the case-specific, excessively fine-grained details have been removed for the sake of generality. A top-level representation of the workflow is shown in Fig. 1. It contains macro-activities or *phases*, numbered from 0 to 4, which will be further described in each sub-process specification, as shown also in Fig. 2-6. Quality checks are spread along the complete workflows, and they must be based on information on the progression and on the outcome of the analysis sub-tasks: the only way to obtain such crucial data is via the inspection of the produced documents. Hence, the quality control system is deeply influenced by the effectiveness of the document management system.

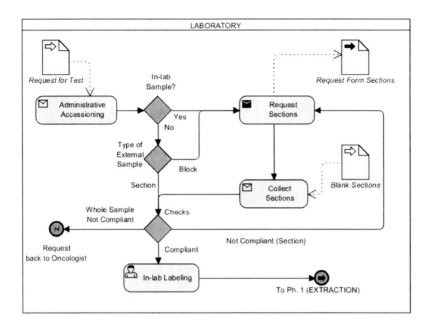

Fig. 2. The check-in phase. Sections can be prepared outside of the laboratory. Compliant sections are labeled for identification in the downstream tasks.

Several variants to the described procedures are possible, and they are mostly due to the nature of the specific test actually performed and to the instrumentation used for it. Despite of the possible multiple variants, the process outline can be considered general enough to keep its validity for any activity of this kind. In particular, it is able to show how to uncover crucial points where semantic tagging is recommendable.

The whole workflow (Fig. 1 and Fig. 2) is cyclically repeated to accommodate the continuously incoming requests. It starts with the reception of a test request, which contains information on what particular kind of test and analysis should be performed, along with patient data and usually also with the disease description. The test request comes with the corresponding biological samples, or sometime with material ready to be processed, as shown in Fig. 2. Typically, any single test cycle operates over a pool of multiple samples from several patients (but not necessarily the same pool throughout the entire workflow). For this reason, whenever it is possible, a preliminary sample sections buffering is implemented. The start of an analysis session is decided by the laboratory supervisor, upon the inspection of the pool of collected requests. This part of the procedure faithfully describes also what happens for investigations related to research issues, although in this case the requests are generated within the laboratory. Definitively, the workflow of a test to be done for research goals does not differ much from the ordinary diagnostic ones. A "core" test session, executed downstream of the check-in phase (Fig. 2), can be thought as composed of four main phases: 1) nucleic acid extraction, 2) amplification, 3) sequencing, and finally 4) reporting.

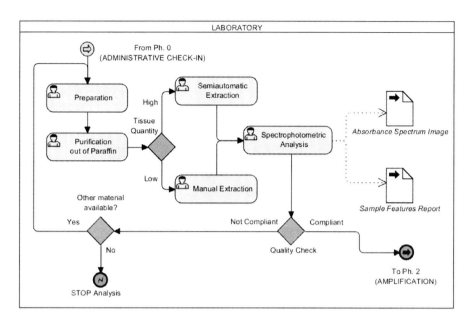

Fig. 3. Representation of the activities in the Nucleic Acid Extraction Phase; two main paths can be followed. The quality check is performed according to the results from a spectrophotometer.

Adjacent phases are separated by quality checks on the results and, in case of unsatisfactory outcomes, the control flow must be redirected backward, in order to possibly recover the procedure. Just to emphasize their role, quality checks are shown both in the overall model (Fig. 1) and in each sub-process (Fig. 2-6). It must be underlined that we take into account two possible technologies to detect DNA sequences/mutations: traditional Sanger sequencing, and pyrosequencing. For this reason, phase 3 has been split in two.

Each phase is carried out making use of specific machinery, and each instrument produces documents, aimed at describing the experiment outcomes and at characterizing the work progression throughout the experiment. The first activity in phase 1 (i.e. nucleic acid extraction, see Fig. 3) is the preparation of samples out of the sections previously obtained. Such preliminary activities are completed taking into account both the patient data and the indications in the test request. Only suitable samples can reach the extraction task. Here, samples are labeled and the binding between patients and samples must be recorded. As part of an intermediate quality check, a spectrophotometer is used, and an image with the absorbance spectrum of the sample is collected from the instrument, as well as a textual report of this subtask. The spectrophotometer results are used to assess the concentration and quality of the extracted nucleic acid.

Phase 2 (i.e. amplification, see Fig. 4) is carried out by means of PCR runs. Depending on the type of sequencing procedure the sample should undergo in the following phase, two alternative paths exist in this sub-process; the use of

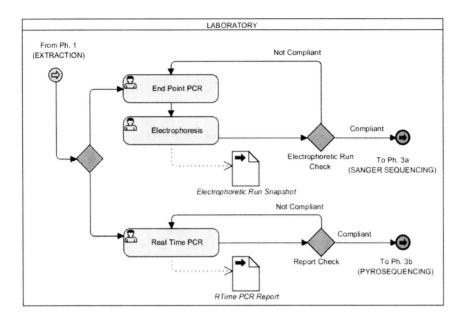

Fig. 4. Diagram for the DNA Amplification Phase. Two distinct paths are followed, depending on the specific procedure to be carried out next (Sanger sequencing or Pyrosequencing).

Real-Time PCR makes it possible to skip an electrophoretic test. Of course, input data for this task describe the DNA segments to be selected (and thus amplified). This information is crucial, and it depends on the test rationale from the requester: research, diagnosis, decisions on pharmacological treatments, etc. As shown in the upper path of Fig. 4, snapshots of gel electrophoreses are generated and collected to assess the outcome of the PCR activity. Typically several electrophoresis runs are present in the same picture, and then one single document aggregates information on multiple samples. Also in this case, it is important to keep trace of the binding between runs and patients. About the generality of the model of this phase, we must underline that the workflow is basically shaped the same way also when we have to work with RNA (in particular with 1-step RT-PCR). Moreover, if cDNA has to be dealt with, an additional retrotranscription task has to be inserted.

In our case, the sequencing phase is made of two alternative sub-tasks: Sanger sequencing and pyrosequencing. Although the employed technologies and the respective outcomes can be considered quite different, the two activities share a common way to proceed, as it is evident in Fig. 5. In the first place, the Sanger sequencing requires a preliminary task for the incorporation of fluorescent dideoxinucleotides and a corresponding purification, while in pyrosequencing a reaction setup activity is present as well; then, the plate setup follows in both cases. Once again, the binding between samples and patients must be recorded, producing an operating document indicated as *sample association matrix*. The

Management of Genotyping-Related Documents   23

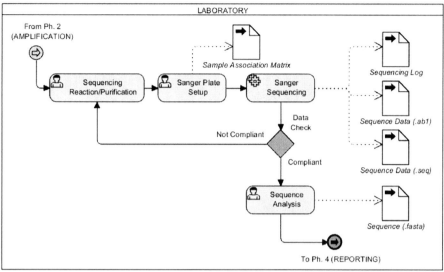

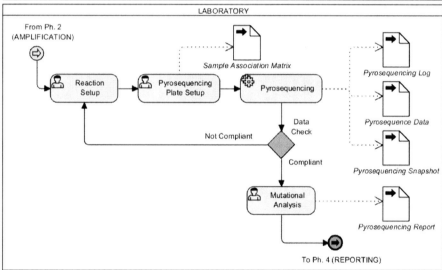

Fig. 5. The Sequencing Phase groups up two distinct tasks that require different instrumentation. Anyway, the arrangement of internal activities is shaped in a similar way.

subsequent sequencing/pyrosequencing run produces documentation on the read sequence/mutation diagrams and on the run progression itself (in Fig. 5 such documents are specified for Sanger sequencing and pyrosequencing). The subsequent inspection of such results, which can be done in a semi-automatic way (because the human intervention is almost always necessary), leads to the determination of the target sequence/mutation, typically outlined in *fasta* format in the classical sequencing. The final phase (see Fig. 6) is aimed at formulating

Table 1. List of documents obtained throughout the genotyping procedures. Depending on the specific document class, a file can correspond to one or multiple biological samples (and thus to one or multiple patients).

Document class	Phase	Content Type	File Format	Patient-Specific/ Aggregated
Sample ID & labeling	Ph. 0	text	.rtf; .doc	PS
Sample ID & labeling data	Ph. 0	data sheet	.xls; .db	A
Abs. spectrum image	Ph. 1	image	.jpg; .tiff; .bmp	PS
Sample features report	Ph. 1	text	.rtf; .doc	PS
Sample features data	Ph. 1	data sheet	.xls; .db	A
RTime PCR report	Ph. 2	PCR log/data	.rex	PS or A
RTime PCR image	Ph. 2	image	.jpg; .tiff; .bmp	PS or A
EF Run Snapshot	Ph. 2	image	.jpg	PS or A
Sequencing log	Ph. 3	run log	.phd.1	PS
Sequence data	Ph. 3	run data	.seq	PS
Sequence-electr	Ph. 3	electropherogram	.ab1	PS
Sequence-text	Ph. 3	text	.fasta	PS
Pyroseq log	Ph. 4	run log	(propr.)	A
Pyroseq data	Ph. 4	run data	(propr.)	A
Pyroseq snapshot	Ph. 4	image	.bmp; .jpg	PS
Pyroseq report	Ph. 4	text	.rtf; .doc	PS
Final report	Ph. 5	text	.rtf; .doc	PS

a precise, concise response from the analysis outcomes. This is achieved via the comparison of the retrieved sequence with the associated known wild type. Such a comparison is aimed at identifying and characterizing possible mutations. A standard report on this finding, written adopting proper standards for mutation descriptions, is filed into a repository that is also accessible from auxiliary personnel, to possibly produce the paper-printed report for the requestor as well.

After a complete description of the workflow phases, we can identify which documents (and their characteristics) are produced along the execution of the tests. Table 1 lists the type of such documents (indicated as "Document class"), the origin phase, the type of the contents, the file format, if the information is related to one or multiple patients/samples, and possible ontologies whose terms could be used to tag them.

4 Management of Genotyping-Related Documents

One of the most significant problems with the described workflow is that all the produced documents are scattered throughout several different, heterogeneous systems aside of the analysis instruments. As a consequence of this kind of physical distribution, it is often difficult to perform cross-platform studies and statistics involving data from multiple phases. An integrated, distributed system for the homogeneous management of the documents involved in the genotyping

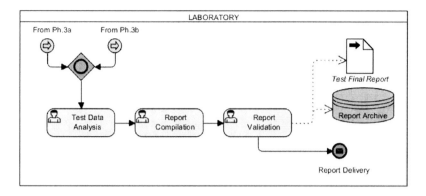

Fig. 6. In the Report Generation Phase, the final test results are presented according to a standard outline

activity would be an ideal solution, and semantic functionalities could be built upon it.

Although integrated software suites for collecting such documents currently exist, they are not widespread at present (especially in small and medium-sized laboratories) because of cost and management issues. From a functional perspective, they can be viewed as typical Document Management Systems (DMS): at the core, a repository containing structured and non-structured documents, and an engine working on top of it and providing features like searching and versioning. Typically DMS are integral parts of Enterprise Content Management (ECM) systems, which provide all the functionalities about security and distribution that are usually required in our target environment. Moreover, integration approaches have been also studied for the development of collaborative, distributed environments in other application fields, and namely e-Business [7]. Therefore, we propose to enhance the organization of this distributed repository by leveraging a technique recently introduced in the document management field: semantic tagging.

4.1 Tagging and Semantic Tagging

The assignment of metadata items to a given resource is usually known as *annotation*. A "resource" may be any kind of information carrier, e.g. a text document, a spreadsheet, a photo, or even an URL. When the involved metadata item is a word, or a short sentence, it is often called *tag* and the corresponding annotation activity is referred to as *tagging*. The intended goal of a tagging system is to describe all resources in a knowledge base in a meaningful way, to ease categorization and future retrieval [39]. Modern ECM systems give the possibility to perform tagging upon the collected documents. The implementation of a tagging system can be based on an underlying relational database that keeps trace of the correspondence of document references/IDs and tags. The actual structure of this supporting tool can affect the performance of the tagging system and the searching procedure, but not the functional characteristics of the ECM. At present, we just focus on functional aspects, leaving possible treatment of performance issues to future works.

In many popular social networks over the Web, tagging is carried out in a collaborative fashion [12], and any user can add tags to any resource. This eventually leads to the emergence of spontaneous bottom-up classifications, known as "folksonomies" [36]. Apart from this social facet, tags are useful even when assigned by a single individual; they are an alternative to traditional folder-based classification, and allow each object to be characterized from multiple viewpoints. However, the categories they introduce are flat, because ordinary tags, unlike folders, have no hierarchical structure. At least, this last statement holds for traditional tagging systems, where users can freely choose the terms to use as tags. Some variants of tagging systems restrict users' choice within a *controlled vocabulary* of some sort; moreover, this vocabulary may contain additional information, such as the relationships among terms (above all, subsumption, or "is-a", which allows the definition of hierarchies). This enhanced vocabulary is usually formalized in an *ontology*.

An ontology is an engineering artefact that is able to formally represent, within a given knowledge domain, all the interesting entities/concepts, their properties and their relationships. The most popular formalism for expressing ontologies is OWL (Web Ontology Language) [6]. As OWL is an XML grammar, it can be proficiently used in different software applications/contexts. *Semantic tagging* refers to precise annotations that can be leveraged in performing a semantic search [39]. Such precise tags can be chosen out of a given ontology, thus obtaining a twofold advantage: first, the meaning of the tag is unambiguously defined; second, relationships between terms in the ontology may be exploited in the search phase to improve the result quality, in terms of both recall and precision metrics. E.g., trivially, if the ontology embeds the information that a "trabecular follicular adenocarcinoma" "is-a" "thyroid neoplasm", we can infer that all queries involving the last term should also consider documents tagged with the more specific term "trabecular follicular adenocarcinoma". Ontologies have been thoroughly investigated in recent literature. In particular bio-ontologies, i.e. those related to biomedical domains, are expected to play more and more important roles in IT support to biomedical investigations [32], and to diagnostic/therapeutic assessments as well. The state-of-the-art in the field of bio-ontologies is relatively advanced, and semantic search engines are becoming more and more available. These two observations can suggest that now favourable conditions are present for the introduction of Semantic Web techniques in the biomolecular test domain, and thus the integration of semantic tagging tools in LIMS becomes a fundamental, strategical step.

4.2 Relying on Standardized Ontologies in the Biomedical Field

So far, the bioinformatics community has been particularly active in the definition and usage of ontologies [11,19,38]: standardized and easily interchangeable domain models are a pressing need when handling extremely large knowledge bases across multiple, closely interacting research groups. The incorporation of bio-ontologies in data annotation systems enables the integration of complex scientific data and supports the consistency of data management activities [11,22]. The OBO initiative (Open Biological and Biomedical Ontology), for example, is a significant classification effort covering many aspects of the biomedical field [35]. It includes various popular projects, most notably the GO (Gene Ontology) [38,5], perhaps the most famous instance of the kind. Started in 1998 with the genomic description of only three organisms, GO now encloses information about genes and gene products of a plethora of animals, plants and microorganisms. The growing usage of GO terms in scientific publications has been monitored [22], and the results of such research point out the crucial role played by this conceptual tool in biology research. Other significant examples of significant ontologies in the biomedical field are Reactome, for biological pathways, Disease Ontology, for human diseases, BioCyc, for both pathways and genomes, and the recently proposed Sequence Ontology (SO), addressing biological sequences.

Biomedical research has pushed the development of various kinds of ontologies, and the presence of concurrent efforts in the field has led to multiple ontology formats, defined independently of the OWL standard or others used in information science: e.g., OBO is a widespread textual representation characterized by low redundancy and high readability; BioPAX (Biological Pathway Exchange) is another example of exchange format, restricted to the domain of biological pathways. The systems biology field is particularly interested by the proliferation of formats [37]. The standardization of ontology languages has subsequently become a compelling need, in order to make easier information exchange. Although minor differences may exist across particular formalisms, the possibility to flexibly use biomedical ontologies in a wider context of software tools has driven a widespread convergence towards OWL [4]. Moreover, also well-developed controlled vocabularies, although their structure would not be so rich as that of typical ontologies, have been recast to OWL: an important example is the NCI Thesaurus, often referred to as NCIT. Several issues in using terms from NCIT for annotation purposes have been reported [25], upon a study on the employment of terms from NCIT to cover all the kinds of entities that are fundamental to an ontology of colon carcinoma. A wider discussion on the semantic model for cancer-related terms in NCIT can be found in [34].

4.3 Applying Tagging to Genotyping Test Procedures

Annotation procedures can be automated in several ways, and often the document contents can be inspected to select possible keywords to be used as metadata. These approaches often are based on sophisticated Natural Language Processing (NLP) techniques that are quite effective. Because of the nature of most of our target documents, such approaches cannot be systematically applied. In fact, the textual content is really limited; Fig. 7 shows an example of the variety of contents of documents from phases 1-3. On the other hand, whenever it is straightforward to extract metadata from a document that is rigidly organized, this operation can be done by using the standard functionalities provided by

28   A. Bechini and R. Giannini

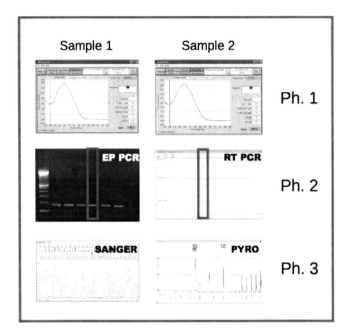

Fig. 7. Some examples of typical contents of documents produced through phases 1-3. The first colum corresponds to a sample that undergoes Sanger sequencing, the other is relative to a pyrosequencing path.

the supporting ECM. A tagging system that supports the whole workflow for genotyping procedures can prove useful at several levels. Apart from these basic observations, ordinary tagging can be used to guarantee document/data lineage retrieval in an integrated way, with no resort to a side relational system. A basic usage of tags may consist in keeping track of all documents related to the same selected sample. Practically, a one-to-one relation between samples and tags has to be stated, and all documents containing results for a given sample can be associated to the related tag. This approach is intended to introduce a uniform means to bind samples (and instruments' runs over them) to the initial request and the patient's data, despite the fact that throughout the whole workflow different binding rules may be actually used in the laboratory practices. This method can be proficiently used in all the workflow points where a "binding" document is produced (see Fig. 2-6).

In the initial request, document tagging with the patient ID can be complemented with a number of semantic tags that describe the actual state of the patient (e.g. diseases) and/or the related biological samples. Here a recourse to terms from Disease Ontology (DOID), or akin, is recommendable. This specific annotation is supposed to be properly propagated at the beginning of phase 1 and beyond, wherever information on request and patient is used. It is not a duty of the tagging tool to enforce the usage of specified ontologies, but instead the laboratory manager is asked to provide internal operating recommendations/guidelines on how to proceed with document annotation.

In the amplification phase, all the documents related to the PCR activities (both end-point and real-time ones) can be obviously annotated with a formal reference to the sequence of nucleotides under investigation. Natural target ontologies for this purpose are GO [5] and more specific ones. As usually one single laboratory may be specialized in genotyping tests in restricted domains, also narrow domain ontologies can be proficiently used. In principle, the effectiveness of the semantic tagging (and especially semantic search) is not jeopardized by the use of very specific domain ontologies, as they can be related to more general and known ones. Anyway, this may lead to relevant problems in actually implementing an efficient search system. Another important issue is determined by the richness and complexity of some well-known ontologies like GO or DOID: their size hampers their management in memory and their graphical representation on the screen, and thus an on-the-fly browsing may be both unpractical for the user and challenging for the annotation tool. A feasible approach is to produce proper subsets of the whole, generic ontologies, containing the terms used in the subdomains covered by the laboratory test targets. Such portions have to contain exactly the original terms, but their modest size make them suitable to be easily handled within the annotation tool. The ontology restriction to a sub-domain can be done either manually or programmatically, according to some predefined criteria: in both cases, the new OWL file with the generated subset will be handled as any other ontology, co-existing with the other loaded ones (the "source" ontology included), and the terms extracted from it will exactly refer to the original concepts. Theoretically, a whole collection of increasingly specific subsets can be kept together in separate OWL files. In this way, if a proper term is not found in a restricted ontology, the user can always resort to a superset of it, even if its handling might be more cumbersome. Moreover, sometimes it is also convenient to have recourse to specific ontologies developed on purpose within the laboratory; in this case, it is advisable to make use of a standard nomenclature.

The general observations done so far also apply to the classification of the gel electrophoresis snapshots (see Fig. 4), and here special care must be paid to the fact that one single document (image) groups up the outcome from multiple samples. As typically such samples relate to the same sequence, investigated for the same reason, this grouping is not particularly relevant for most of the classification terms.

The output documents from the DNA sequencing runs (Fig. 5) are the description of the base-specific intensity profiles (in proprietary and/or open formats), and the log of the sequencing activity. Tags here might also be aimed at pointing out possible problems in the interpretation of the intensity profiles. Again, the obtained *fasta* sequence must be precisely related to the studied exact sequence (but this information can be obtained automatically from previous annotations on the samples worked out).

The final diagnosis is suitable to be tagged with standard notations for the mutations possibly found in the sample [29]. Although here no ontological information is apparently embedded, the mutation code can be related (out of the context of the used software framework) to other ontological models associated to the mutation. Even this last kind of annotation can further enable an ontology-assisted search of the collected knowledge base.

5 Management and Classification within an ECM

The proposed solution for the integration of management and annotation tasks for the target documents is based on a standard document management system (DMS), which typically comes as a core component of a larger Enterprise Content Management (ECM) suite. In other words, a properly customized ECM can be used as the backbone of a LIMS for supporting the genotyping process. Currently, ECMs support metadata handling in various fashions, but not semantic tagging, at least in the form described in this paper. This means that the chosen ECM must be enhanced with the semantic tagging functionality, which should complement those already present. In fact, the ordinary free-tag annotation misses a couple of crucial benefits: i) annotation precision/unambiguity, and ii) possibility to exploit the knowledge embedded in the dense network of relations present among ontology terms. Of course, the annotation support must avoid that a free keyword coud be mistaken for an homonymous term out of a formal ontology. Such a semantic tagging support has then been designed and developed for the Alfresco ECM, and further details on it can be found in subsection 5.2.

An ECM is also a good candidate to act as the LIMS core component because it can be equipped to be accessible from a heterogeneous set of coworkers, and to easily interoperate with external applications, e.g. through APIs and/or Web Services. We expect that both the quality and the accessibility of information in repositories for genotyping documents would benefit from the employment of a single, solid, and easily-accessible software system.

5.1 Organizing an ECM as a Laboratory Information Management System

Our proposal for a Laboratory Information Management System relies on a properly shaped DMS, whose repository contains all the documents that get created during analyses over samples. To access the repository, we can plainly leverage the DMS user interface. Modern DMSs are web-based applications, accessible (upon authentication) from distributed web clients; we assume that, using web clients from all the computers aside the laboratory instruments, it is possible to upload the output documents directly into the DMS repository. If this operation cannot be done in place, document files must be moved somehow to computers in the neighbourhood to perform the required uploads.

Our prototype is based on Alfresco ECM [1], a popular open source Enterprise Content Management system. The Alfresco framework integrates modules devoted to Document Management and Web Content Management. It is a web-based application that relies on Java technologies and also makes widespread use of JavaScript components. Its modular architecture makes it suitable to be customized and specifically shaped to accommodate our requirements. In Alfresco, the principal means to organize documents is via *spaces*, a concept similar to that of folders, but in a web-accessible setting; spaces can be managed by multiple collaborating actors working through different web clients. Spaces can be characterized also at a behavioral level and, specifically, *rules* can be added to manage content being inserted, edited in, or removed from a space. One of the actions a rule may trigger is the activation of a functionality, or *aspect*, for documents in the space that satisfy a given condition; typical aspects in Alfresco are known as "versionable", "taggable" or "classifiable".

Spaces can be used in the LIMS to provide a guiding structure to the document management and annotation tasks. The idea is to simply apply the classical folder-based classification in the archiving job, to avoid to waste time in attaching tags with "ordinary" values. Thus in the first place the operator is guided to perform a folder-based classification for the uploaded file. In other words, the document has to be placed in a specific folder aimed at collecting all the material which must be annotated according to a precise term. So, a hierarchy of spaces must be defined to receive the uploaded files; we will call it *host hierarchy*. It mimics a tree-shaped classification for certain criteria. E.g., each first-level space may correspond to the anatomical entity the related sample originates from, the second-level to a specific test requested on the sample, and so on. Automatic procedures can be triggered by the insertion in such spaces so that precise tag values would be attached to the document; this way, even if the file will be successively moved, this specific classification would not be lost. Practically, the initial folder-based classification induces the automatic annotation for the most straightforward terms. After this initial basic step, the user must be able to annotate the document with other multiple tags, and we advocate the opportunity to use semantic ones. Unfortunately Alfresco, in its current basic configuration, supports only ordinary tagging, not the semantic variant. For this reason, we must implement a modification to enable ontology management.

5.2 Seamless Integration of Semantic Tagging

The semantic tagging functionality can be introduced in Alfresco in a couple of different ways:

- modifying the behaviour related to handling a specific aspect, namely "taggable"
- adding a new aspect (say "semantic taggable") and providing proper handling procedures for it.

Regardless of the choice, which just marginally impacts the actual steps to be followed in performing the operation, we have to decide how to specify and to handle the ontologies we want to consider in our simple semantic-aware environment. The ontology contents must be accessible to the framework and, although

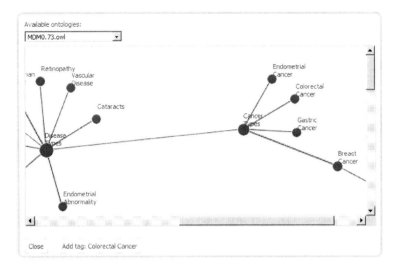

Fig. 8. A graphical layout of the target ontology can be used to help in finding the most correct term to add. In the reported screenshot, client-side JavaScript code is used to display ontology terms as a force-directed graph.

this information could be even obtained remotely, a local copy of the file makes it promptly available. A flexible way to deal with ontologies is to assign them a dedicated place (a *space*) in the DMS repository, where OWL files can be freely uploaded (or dropped). During this procedure, they are handled just like any other document in the DMS and no specific action has to be performed.

The space where the target documents must be initially uploaded is required to hold the aspect used for the introduction of semantic tags (for the sake of simplicity, we suppose it would be the "taggable" one). We must make sure to add a rule to all the spaces in the host hierarchy such that each incoming document will become "taggable", i.e. extended with the "taggable" aspect. The rationale for setting this requirement is the practical need to guarantee that tagging would always be enabled for the documents to be classified.

The ECM web user interface heavily relies on JavaScript modules. This structural feature lets us implement a lightweight support to ontology-aware tagging, which is not intrusive for the whole DMS framework because it requires only some modifications to the behavior of the tagging module. Following this low-impact philosophy, we chose a JavaScript-only approach to both ontology processing and visualization, acting just at the presentation layer of the web application. The starting point for adding a semantic tag is the same as for ordinary tagging, i.e. in the portion of the web interface dedicated to the management of the document properties (known as the "property view" pop-up). This GUI component can be enriched so that, in case of a "taggable" document, a special link (labelled as "Add a tag") would appear to provide access to a new navigation interface. In the first place, the activated graphical procedure lets the user choose a target ontology; this can be naturally accomplished by means of a combo-box that reports the list of OWL files uploaded into the dedicated repository space for possible target ontologies. Upon such a selection, a graphical representation for the chosen ontology it is shown in the web user interface, and it can be interactively explored for an easy identification of interesting concepts. To this aim, we made use of a force-directed graph layout for an animated visualization, where nodes are ontology concepts and links between nodes represent a parent-child relation, as shown in Fig. 8. Interactive navigation is performed by acting upon the nodes that, once clicked, reveal direct children (if present). Moreover, upon clicking the node, the corresponding term is selected as a candidate value and the text on the button for attaching the new tag is accordingly changed. Of course, other representations could be proposed to make the graphical navigation more immediate; in any case, the possibility to show entire graph portions clustered into one single node is fundamental to handle very rich ontologies.

Finally, when the user has found the entity/concept of interest, he can add the entity name to the document's tag set. In order to avoid ambiguities, the resulting tag will include both the name of the entity and a reference to the specific target ontology. As we will see in the subsequent section, it is particularly useful to keep a list of the most recently used tag values, to ease the completion of the subsequent annotations upon other documents.

From our experience we can undoubtedly state that the implementation in Alfresco of the proposed solution is painless from both the development and the deployment perspectives. In practice, only modifications at the presentation level are sufficient, and mostly they just involve JavaScript code. If we decide to add a specific custom aspect, the new components to be inserted partially span the server-side tiers of the system, but do not interfer with the other modules.

6 Case Study: Genotyping for Pharmacogenetics-Based Therapeutic Recommendations

The employment of an ECM as the core component of a LIMS and the described semantic tagging procedures have been evaluated taking the daily activity of a medium-sized genotyping laboratory as a reference. The core activity of our target laboratory is the execution of diagnostic tests on samples of neoplastic tissues, to uncover mutations that affect therapeutic treatments. The genotyping result is crucial for the oncologist to give pharmacogenetics-based therapeutic recommendations. Every procedure must be carried out with particular care, and stringent quality control practices are applied. According to the proposed approach, all the sample-related documents obtained throughout the genotyping procedures must be tagged with an appropriate, standard description term. Specifically, semantic annotations have to describe unambiguously the key characteristics of the samples, i.e., the anatomical entity, the histotype of neoplasm, the analyzed gene or genes and the obtained results (mutations nomenclature or description). Additionally, other tags can be adopted to describe the used methods and analytical data for every sample. We can briefly list the main types of annotations required in this context, pinpointing and related ontologies useful in our target case:

- tumor type by anatomical entity (i.e., possible values from DOID are "malignant neoplasm of lung", "malignant neoplasm of thyroid", etc.);
- annotation of the tumor histotypes (i.e. "mucinous adenocarcinoma of the rectum", "non small cell lung cancer", etc.);
- name of the gene of interest (i.e., "BRAF", "Kras", "Hras", "EGFR", etc.);
- specification of the analyzed exon/exons (i.e., "BRAF Exon-15", "EGFR Exon-18", "Kras Exon-2", etc.);
- specification of the mutation/SNP possibly found (i.e., using the *Mutation Nomenclature*, "BRAF (Ex. 15) c.1799T>A p.V600E", "Kras (Ex.12) c.35G>T p.G12V", etc.).

We should pinpoint that, as already discussed, several standard ontologies exist to meet the presented requirements, and it is up to the laboratory direction to decide what specific ones to consider for the internal practices.

A simple test was designed to comprehend whether the overall information management organization would be suitable for the target setting, and to quantify the overhead introduced by the application of the semantic tagging procedures with respect to an ordinary workflow. The outcomes are summarized in Table 2.

The used dataset was pertinent to the main work carried out in the laboratory: the test dataset was made of the whole collection of documents stored in the computers aside of the instruments used in each phase of the genotyping procedure, and generated during more than one year of activity. Different document categories have been considered, to understand whether the document type may somehow influence the results. The host hierarchy was organized according to the involved anatomical entity on the first level, and according to the test types classification on the underlying ones. To keep their size manageable, the used ontologies were taken from subsets of larger ones, and one of them was developed on purpose.

Table 2. Timing characterization of the tagging procedure. The results on the upper part refer to learning operators; on the lower, values refer to the same operators after a brief training period.

	D1	*D2*	*D3*	*D4*
Upload time (avg, secs)	27	20	25	24
Tagging time (avg per tag, secs)	32	35	28	29
% tags picked on graph	100%	100%	100%	100%
% tags from "most used" list	0%	0%	0%	0%
Upload time (avg, secs)	15	18	17	13
Tagging time (avg per tag, secs)	12	9	14	11
% tags picked on graph	25%	20%	35%	25%
% tags from "most used" list	75%	80%	65%	75%

Legend
D1: Abs. spectrum image; D2: Sample Features Report; D3: EF Run Report; D4: Pyroseq. Report

The ECM interface was presented to the laboratory personnel, that actually are domain experts. In a first phase, it was asked to upload and tag files belonging to four different document classes, randomly picked up out of the dataset. It must be underlined that the inspected uploading operation was supposed to start with the choice of the Alfresco space where to add the document. In this occasion, they have been constrained to always explore the graphical ontology representation, with no resort to the "most used" tag list. The conditions and contraints set in this first phase were aimed at obtaining a sensible estimate of the upper bound for the operating times, and the results are shown in the first part of Table 2, organized in separate columns according to the involved document type.

Later, after a brief training period to get accustomed with the tool, the same people were asked to perform exactly the same operations. This second phase was aimed at obtaining the average timings that can reasonably be expected during the ordinary execution of the genotyping procedures.

By inspecting the results in Table 2, no significant difference in timings across document classes becomes evident. By comparing the values obtained in the two phases, we can see that likely the operators are initially much more familiar with the folder-based classification, but soon the operation of adding a tag becomes much quicker and more efficient (this might be due to the availability of frequently used tags; in folder-based classification, the host hierarchy must always be navigated to select the proper space). Although it cannot be directly ascertained from the values in Table 2, we think that a thoughtful, balanced organization of the host hierarchy is decisive to improve the tagging timings, because a preliminary choice of the destination space implicitly cut off most of the ontologies as not applicable, and thus the subsequent selection of the target ontology becomes easier and quicker. The high percentage of tag reuse (reported in the last row) is obviously determined by the fact that each laboratory operates on a restricted number of domains; in any case, such values can be reasonably obtained whenever operating over a batch of tests within the same domain.

The described experience is necessarily limited, because we can rely only on domain experts within the lab (it makes no sense to ask someone to use a domain ontology whose terms are totally meaningless for him). Anyway, it clearly shows that semantic tagging is not excessively time-consuming, especially if compared with traditional forms of document classification and archiving. The combined effect of the host hierarchy and the semantic tagging seems to make the classification more effective and efficient. Furthermore, the time spent in tagging is negligible within the overall genotyping test duration (it may take up to 5-6 days), and the overall laboratory throughput should not be affected. On the other hand, the effectiveness of quality control procedures definitely benefits from a precise classification of documents.

Tag-based search is a standard functionality supported by Alfresco, and it can be used referring to the attached terms, regardless of the fact that they belong to a certain ontology. This simple approach misses most of the benefits of semantic search, because it does not take into account the knowledge embedded in the referred ontology. Recent works have tried to depict both qualitatively and quantitatively the benefits of semantic search [31], especially in specific domains. In our case, a simple extension of the plain tag-based search towards semantic awareness can be done by modifying the initial set of tag values in a query, exploiting information from an external reasoner on concepts related to the initial terms. At this point, it is difficult to carefully quantify to what extent this way to proceed could be beneficial to the end-user; in general, about the effectiveness of semantic search in terms of document retrieval metrics, it is a common belief that further experimentation would be needed [16].

7 Conclusions

The application of semantic approaches to shape the way biomedical data are handled and exploited can be recognized as a growing trend in the last years. Such an approach can be pushed further, and also integrated in a seamless way with an overall framework for the management of documents produced during the laboratory activities.

In this paper, stemming from experience on the development of a prototype system targeted to the management and annotation of genotyping-related documentation, an ECM is proposed as the main component of a semantically-enabled LIMS. Such solution employs a traditional DMS, and the required modifications can be easily implemented, because they do not interfere with the core system components. From a functional standpoint, it is crucial to leave the user free to choose any possible ontology (both upper- and domain ones) for selecting precise terms for document annotation, and this requirement has been well accommodated. Moreover, ordinary tagging can go along with semantic one, and the flexibility of the tagging tool makes it usable in the context of different operating procedures adopted within a laboratory. Nonetheless, once a semantically annotated document base would be available, it can be exploited also with the help of the knowledge embedded in the used ontologies, possibly making use of specific external reasoning engines. The availability of open source ECMs can represent another key factor for the success of the proposed approach, paired to the fact that experimentation has shown that it does not introduce significant overhead in the ordinary test progression workflow. We thus believe that the adoption of a well-structured, semantics-enabled document management system like the one described in this paper could effectively contribute to the day-to-day operations biologists are involved in, building up an ever-growing, valuable, semantically searchable knowledge base.

Future developments will try to put in place mechanisms in the ECM to make the internal tagged document base available to external systems, which can resort to effective reasoners to fully exploit the information potential inside the document repository. Such mechanisms are also expected to lay the foundations for a possible federation of semantic document bases from single separated laboratories.

Acknowledgments. The main acknoledgement is due to Jacopo Viotto, who collaborated to the first version of the described system and to a previous paper on the topic. Alessandro Spagnesi has provided the first raw drafts for the BPMN diagrams. Special thanks for their kind collaboration are due to the personnel of the biomolecular analysis laboratory of the div. of Anatomical Pathology IV in Azienda Ospedaliera Pisana, in Pisa, Italy.

References

1. Alfresco website, http://www.alfresco.com
2. MeSH - Medical Subject Headings, http://www.nlm.nih.gov/mesh/
3. Altman, R.B., Bada, M., Chai, X.J., Carrillo, M.W., Chen, R.O., Abernethy, N.F.: RiboWeb: An ontology-based system for collaborative molecular biology. IEEE Intelligent Systems 14(5), 68–76 (1999)
4. Aranguren, M.E., Bechhofer, S., Lord, P., Sattler, U., Stevens, R.: Understanding and using the meaning of statements in a bio-ontology: recasting the Gene Ontology in OWL. BMC Bioinformatics 8, 57 (2007)
5. Ashburner, M., Ball, C., Blake, J., Botstein, D., Butler, H., Cherry, M., Davis, A., Dolinski, K., Dwight, S., Eppig, J.: Gene Ontology: Tool for the unification of biology. Nature Genetics 25, 25–29 (2000)
6. Bechhofer, S., van Harmele, F., Hedler, J., et al.: OWL Web Ontology Language reference (2002)
7. Bechini, A., Tomasi, A., Viotto, J.: Collaborative e-business and document management: Integration of legacy DMSs with the ebXML environment. In: Interdisciplinary Aspects of Information Systems Studies, pp. 287–293. Physica-Verlag HD, Heidelberg (2008)
8. Bechini, A., Tomasi, A., Viotto, J.: Enabling ontology-based document classification and management in ebXML registries. In: Proceedings of ACM SAC, pp. 1145–1150. ACM, New York (2008)
9. Bechini, A., Viotto, J., Giannini, R.: Smooth introduction of semantic tagging in genotyping procedures. In: Khuri, S., Lhotská, L., Pisanti, N. (eds.) ITBAM 2010. LNCS, vol. 6266, pp. 201–214. Springer, Heidelberg (2010)
10. Berners-Lee, T., Hendler, J., Lassila, O.: The semantic web. Scientific American 284(5), 34–43 (2001)
11. Bleke, J.: Bio-ontologies - fast and furious. Nature Biotechnologies 6(22), 773–774 (2004)
12. Bojars, U., Breslin, J.G., Peristeras, V., Tummarello, G., Decker, S.: Interlinking the social web with semantics. IEEE Intelligent Systems 23(3), 29–40 (2008)
13. Choy, D., Brown, A., McVeigh, R., Müller, F.: OASIS Content Management Interoperability Services (CMIS) Version 1.0 (2010)
14. Deus, H.F., Stanislaus, R., Veiga, D.F., Behrens, C., Wistuba, I.I., Minna, J.D., Garner, H.R., Swisher, S.G., Roth, J.A., Correa, A.M., Broom, B., Coombes, K., Chang, A., Vogel, L.H., Almeida, J.S.: A semantic web management model for integrative biomedical informatics. PLoS ONE 3(8), e2946 (2008)
15. Ding, L., Finin, T.W., Joshi, A., Peng, Y., Pan, R., Reddivari, P.: Search on the semantic web. IEEE Computer 38(10), 62–69 (2005)
16. Dong, H., Hussain, F.K., Chang, E.: A survey in semantic search technologies. In: Proc. of DEST 2008, 2nd IEEE Int'l Conf. on Digital Ecosystems and Technologies, pp. 403–408 (2008)

17. Donofrio, N., Rajagopalon, R., Brown, D.E., Diener, S.E., Windham, D., Nolin, S., Floyd, A., Mitchell, T.K., Galadima, N., Tucker, S., Orbach, M.J., Patel, G., Farman, M.L., Pampanwar, V., Soderlund, C., Lee, Y.-H., Dean, R.A.: 'paclims': A component LIM system for high-throughput functional genomic analysis. BMC Bioinformatics 6, 94 (2005)
18. Fong, C., Ko, D.C., Wasnick, M., Radey, M., Miller, S.I., Brittnacher, M.J.: Gwas analyzer: integrating genotype, phenotype and public annotation data for genome-wide association study analysis. Bioinformatics 26(4), 560–564 (2010)
19. Hadzic, M., Chang, E.: Medical ontologies to support human disease research and control. International Journal of Web and Grid Services 1(2), 139–150 (2005)
20. Huang, Y.W., Arkin, A.P., Chandonia, J.-M.: WIST: toolkit for rapid, customized LIMS development. Bioinformatics 27(3), 437–438 (2011)
21. Jayashree, B., Reddy, P.T., Leeladevi, Y., Crouch, J.H., Mahalakshmi, V., Buhariwalla, H.K., Eshwar, K.E., Mace, E., Folksterma, R., Senthilvel, S., Varshney, R.K., Seetha, K., Rajalakshmi, R., Prasanth, V.P., Chandra, S., Swarupa, L., SriKalyani, P., Hoisington, D.A.: Laboratory information management software for genotyping workflows: applications in high throughput crop genotyping. BMC Bioinformatics 7, 383 (2006)
22. Jensen, L.J., Bork, P.: Ontologies in quantitative biology: A basis for comparison, integration, and discovery. PLoS Biology 8(5), e1000374 (2010)
23. Kohl, K., Gremmels, J.: Documentation system for plant transformation service and research. Plant Methods 6(1), 4 (2010)
24. Kothari, C.R., Wilkinson, M.: Structured representation of biomedical experiments: A bottom-up approach. In: Proceedings of Int'l Conf. on Information and Knowledge Engineering (IKE), pp. 199–204. CSREA Press (2008)
25. Kumar, A., Smith, B.: Oncology ontology in the NCI thesaurus. In: Miksch, S., Hunter, J., Keravnou, E.T. (eds.) AIME 2005. LNCS (LNAI), vol. 3581, pp. 213–220. Springer, Heidelberg (2005)
26. Le Hellard, S., Ballereau, S.J., Visscher, P.M., Torrance, H.S., Pinson, J., Morris, S.W., Thomson, M.L., Semple, C.A.M., Muir, W.J., Blackwood, D.H.R., Porteous, D.J., Evans, K.L.: SNP genotyping on pooled DNAs: comparison of genotyping technologies and a semi automated method for data storage and analysis. Nucleic Acids Research 30(15), e74 (2002)
27. Li, J.-L., Deng, H., Lai, D.-B., Xu, F., Chen, J., Gao, G., Recker, R.R., Deng, H.-W.: Toward high-throughput genotyping: Dynamic and automatic software for manipulating large-scale genotype data using fluorescently labeled dinucleotide markers. Genome Res. 11(7), 1304–1314 (2001)
28. Monnier, S., Cox, D.G., Albion, T., Canzian, F.: T.I.M.S: TaqMan Information Management System, tools to organize data flow in a genotyping laboratory. BMC Bioinformatics 6, 246 (2005)
29. Olivier, M., Petitejan, A., Teague, J., Forbes, S., Dunnick, J., der Dunnen, J., Langerod, A., Wilkinson, J., Vihinen, M., Cotton, R., Hainaut, P.: Somatic mutation databases as tools for molecular epidemiology and molecular pathology of cancer: Proposed guidelines for improving data collection, distribution, and integration. Human Mutation 30(3), 275–282 (2009)
30. OMG. BPMN 2.0 specifications (2009)
31. Price, S.L., Nielsen, M.L., Delcambre, L.M., Vedsted, P., Steinhauer, J.: Using semantic components to search for domain-specific documents: An evaluation from the system perspective and the user perspective. Information Systems 34(8), 724–752 (2009)

32. Rubin, D.L., Shah, N.H., Noy, N.F.: Biomedical ontologies: a functional perspective. Briefings in Bioinformatics 9(1), 75–90 (2008)
33. Shah, N., Jonquet, C., Chiang, A., Butte, A., Chen, R., Musen, M.: Ontology-driven indexing of public datasets for translational bioinformatics. BMC Bioinformatics 10(suppl.2), S1 (2009)
34. Sioutos, N., de Coronado, S., Haber, M.W., Hartel, F.W., Shaiu, W.L., Wright, L.W.: NCI Thesaurus: a semantic model integrating cancer-related clinical and molecular information. Journal of Biomedical Informatics 40(1), 30–43 (2007)
35. Smith, B., Ashburner, M., Rosse, C., Bard, J., Bug, W., Ceusters, W., Goldberg, L.J., Eilbeck, K., Ireland, A., Mungall, C.J., Consortium, T.O., Leontis, N., Rocca-Serra, P., Ruttenberg, A., Sansone, S.-A., Scheuermann, R.H., Shah, N., Whetzel, P.L., Lewis, S.: The OBO foundry: coordinated evolution of ontologies to support biomedical data integration. Nature Biotechnology 25, 1251–1255 (2007)
36. Specia, L., Motta, E.: Integrating Folksonomies with the Semantic Web. In: Franconi, E., Kifer, M., May, W. (eds.) ESWC 2007. LNCS, vol. 4519, pp. 624–639. Springer, Heidelberg (2007)
37. Strömbäck, L., Hall, D., Lambrix, P.: A review of standards for data exchange within systems biology. Proteomics 7(6), 857–867 (2007)
38. Tanabe, L.K., Wilbur, W.J.: Tagging gene and protein names in biomedical text. Bioinformatics 18(8), 1124–1132 (2002)
39. Uren, V., Cimiano, P., Iria, J., Handschuh, S., Vargas-Vera, M., Motta, E., Ciravegna, F.: Semantic annotation for knowledge management: Requirements and a survey of the state of the art. Journal of Web Semantics 4(1), 14–28 (2006)
40. Wohed, P., van der Aalst, W.M.P., Dumas, M., ter Hofstede, A.H.M., Russell, N.: On the suitability of BPMN for business process modelling. In: Dustdar, S., Fiadeiro, J.L., Sheth, A.P. (eds.) BPM 2006. LNCS, vol. 4102, pp. 161–176. Springer, Heidelberg (2006)

MEDCollector: Multisource Epidemic Data Collector

João Zamite[1], Fabrício A.B. Silva[2], Francisco Couto[1], and Mário J. Silva[1]

[1] LaSIGE, Faculty of Science
University of Lisbon, Portugal
[2] Army Technology Center
Information Technology Division
Rio de Janeiro, Brazil
epiwork@lasige.di.fc.ul.pt

Abstract. We present a novel approach for epidemic data collection and integration based on the principles of interoperability and modularity. Accurate and timely epidemic models require large, fresh datasets. The World Wide Web, due to its explosion in data availability, represents a valuable source for epidemiological datasets. From an e-science perspective, collected data can be shared across multiple applications to enable the creation of dynamic platforms to extract knowledge from these datasets. Our approach, MEDCollector, addresses this problem by enabling data collection from multiple sources and its upload to the repository of an epidemic research information platform. Enabling the flexible use and configuration of services through workflow definition, MEDCollector is adaptable to multiple Web sources. Identified disease and location entities are mapped to ontologies, not only guaranteeing the consistency within gathered datasets but also allowing the exploration of relations between the mapped entities. MEDCollector retrieves data from the web and enables its packaging for later use in epidemic modeling tools.

Keywords: Epidemic Surveillance, Data Collection, Information Integration, Workflow Design.

1 Introduction

The study of epidemic disease propagation and its control is highly dependent on the availability of reliable epidemic data. Epidemic surveillance systems play an important role in this subject, extracting exhaustive information with the purpose of understanding disease propagation and evaluating its impact in public health through epidemic forecasting tools.

International organizations, such as the World Health Organization (WHO), have epidemic surveillance systems that collect infectious disease cases. However, although official disease statistics and demographics provide the most reliable data, the use of new technologies for epidemic data collection is useful to complement data already obtained from national reporting systems.

In recent years, several projects have researched the use of the Web as a platform for epidemic data collection. The systems developed by these projects gather epidemic data from several types of sources [1], such as query data from search engines [2], Internet news services[3] and directly from users [4]. Alternative sources for epidemic data are social networks, e.g. Twitter [5], which are forums where people share information that can be accessed as Web services. These alternative sources of information can be used to identify possible disease cases, or at least provide a glimpse about the propagation of a disease in a community.

The aforementioned systems, based on Web technologies, contribute to the recent increase in available epidemic data. However, this data is neither centralized nor organized in order to be easily found and shared between scientists and health professionals. Furthermore, these systems use different protocols for data presentation and transmission and as such there is no unified way to extract and use all of this available data. Therefore, an integrative effort is required to consolidate this data so it can be used in e-science data-analysis.

The Epidemic Marketplace (EM) [6] developed by the EPIWORK project is an information platform for the integration, management and sharing of epidemiological data. EM stores data derived from Internet monitoring systems and is designed to interoperate with a computational modeling platform. It also provides a forum for the discussion of epidemic datasets and epidemic modeling issues. This framework comprised of information and computations platforms will provide tools for data management, epidemiological modeling, forecasting and monitoring (Fig. 1).

The EM is composed of several different modules which provide the functionalities of the information platform. These include a digital repository of epidemic datasets and other resources and mediator services which enable applications to interact with the digital repository. To tackle the aforementioned necessity of data integration from multiple sources the EM also includes the MEDCollector, a workflow design system for the extraction and integration of epidemic data from multiple Web sources (Fig. 2).

MEDCollector follows a workflow design approach to the extraction and integration of data from multiple heterogeneous Web sources. Workflow definitions enable the declaration of collection mechanisms using web services (Fig. 3). Through workflows extraction mechanisms can be flexible, allowing the addition of new functionalities, through services, changing the way these mechanisms behave.

Since workflow design tools are typically complex and require users to have specific technical workflow knowledge, such as business analysts, there is the need for the creation of abstractions and simplifications to the process of creating workflows. This can be achieved through a simple workflow design interface which enable non-technical users to create the necessary data harvesting mechanisms.

Following this approach, this paper describes the development of MED Collector, a system for information extraction and integration from multiple

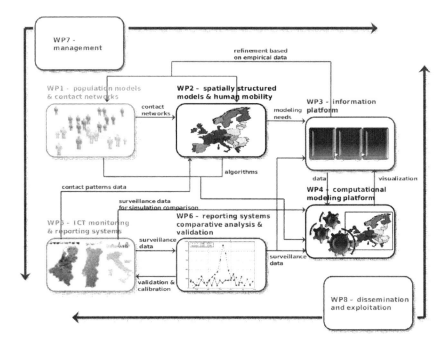

Fig. 1. The main components of the EPIWORK project and how they interact to study epidemics. Figure extracted from http://www.epiwork.eu/

heterogeneous epidemiological data sources using worklows. MEDCollector enables the flexible configuration of epidemic data collection from multiple sources, using interoperable services orchestrated as workflows. Collected data can then be packed into datasets and uploaded to EM's repository for later use by epidemic modeling tools. Through the use of Web standards for data transmission, the system can be seamlessly integrated with external web services to extend its basic functionality. This system gathers and integrates data from multiple and heterogeneous sources, providing epidemiologists a wide array of datasets obtained from the Web using its own data services, in addition to traditional data sources.

This paper is an extended version of the conference paper with the same title presented at ITBAM'10 [7]. This paper extends MEDCollector to accomplish the requirements of data provenance and packaging and also includes an evaluation of the system by comparison with other scientific workflow systems.

The remainder of the paper is organized as follows: Section 2 provides insight into previous related work; Section 3 is an assessment of the system requirements for an epidemic data collector; Section 4 presents our system's architecture and a brief description of its implementation; Section 5 compares MEDCollector with two other workflow design software systems, Taverna and Kepler; Section 6 presents the conclusions and perspectives for future work in MEDCollector.

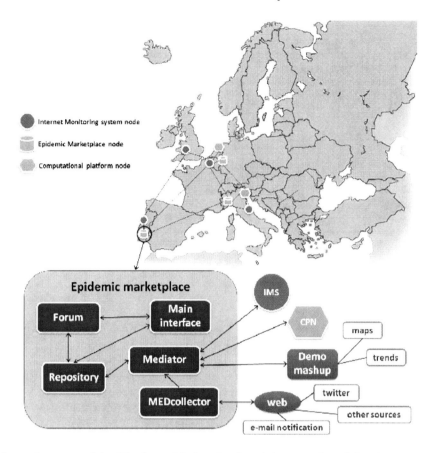

Fig. 2. Overview of the "Epidemic Marketplace" and the several modules that compose it

2 Related Work

Over the past few years, with the explosion of the World Wide Web, online data availability has increased significantly. With this increase, news systems for browsing, harvesting and retrieving this information have surfaced. These systems have taken advantage not only of the topology of the Web but also from the large user base that it represents. For instance, DROP [8], a system for harvesting and storing online publications, not only includes a crawler module that collects publications actively from the web, but also accepts direct publication submissions to the system.

The development of middleware and networking technologies for tasks such as data acquisition, integration, storage, management, mining and visualization enables the use of web, computational and information technologies to achieve scientific results, or E-science. E-science, through the provision of scientific environments, allows global collaboration by enabling universal access to knowledge and resources.

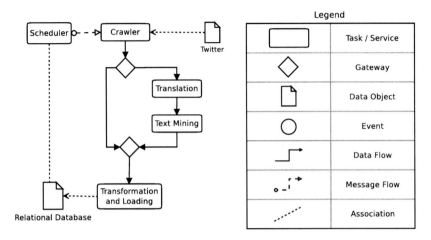

Fig. 3. Example of a workflow to extract messages from twitter, text-mine them, and insert the extracted information into a database

Taverna, developed by the myGrid team [9] [10] is an open source Workflow Management System commonly used in bioinformatics which allows users to design and execute scientific workflows. Taverna enacts workflows as a series of connected web services, enabling the creation of *in silico* experiments with data available on the Web.

Kepler [11] is also an open source scientific workflow system. However, unlike Taverna, Kepler's functionalities are not dependent on web services, but on actors which do not depend on Web Servers.

Together with Tavera and Kepler, other initiatives such as EGEE and DEISA [12] are bridging the gap between the need for computation tools and their seamless integration through the use of standards and interoperable software.

The use of workflow environments enable the creation of mechanisms to transform and structure and integrate data from multiple sources, originally in different formats and dealing with multiple concepts, through adequate services which can be orchestrated for that purpose.

The Web presents a valuable source for collecting epidemic data, but it requires coping with a variety of formats, ranging from free text to XML documents. Disease reporting services, like the ProMED-mail newsletter [13], EuroFlu and reports from the European Centre for Disease Prevention and Control (ECDC) [14] are useful sources of epidemiological data. The ProMed-mail newsletter, maintained by the International Society for Infectious Diseases, is a notification service that sends their registered users information about new disease outbreaks and cases via e-mail. EuroFlu.org, a WHO website, and the ECDC's European Influenza Surveillance Network (EISN) [15] publish weekly reports on the activity of Influenza-like diseases.

Internet Monitoring Systems (IMS) can retrieve data using two distinct approaches: passive data collection and active data collection. Systems that use passive data collection mechanisms, such as Gripenet [4] and Google Flu Trends [2], provide interfaces for their users to voluntarily submit their data. On the other hand, active data collection systems, such as Healthmap [1] and the Global Public Health Intelligence Network (GPHIN) [3], use crawlers that browse the Web through hyper-links and available Application Programming Interfaces (API).

Gripenet is a Portuguese IMS which depends directly on the active participation of its voluntary users, which receive weekly newsletters about influenza activity and are requested to fill out a form about the presence, or not, of influenza symptoms during the past week. This system was based on Holland's Influenzanet [16] model and is currently implemented on seven other countries: Belgium, Italy, Brazil, Mexico, United Kingdom and Australia and Canada.

Google Flu Trends is a system that uses search terms from user queries to the Google search engine to estimate influenza activity each week. Google Flu Trends has been shown to predict influenza activity within two weeks prior to the official sources for the North American population. This system is currently being extended to cover other countries around the world. Both Google Flu Trends and the previously mentioned IMS collect data directly from their users; their difference lies in that while Google Flu Trends receives data directly from its users this data is not directly related to disease activity. Google Flu Trends correlates a number of search terms with disease activity increase and then uses them to estimate real activity values.

Healthmap [1] takes a different approach. It is a worldwide epidemic data presentation website that depicts disease cases, mostly of contagious diseases, gathered from different sources. These sources can be diverse in nature, ranging from news casting services to official epidemic reports, and have different degrees of reliability. Disease and location information is extracted via a text processing system and presented on a map via the Google Maps API.

GPHIN [3] is a multilingual early warning system for global public health surveillance. It gathers reports of public health significance from news-feed aggregators. GPHIN then filters and categorizes the retrieved reports through an automated process which is complemented by human analysis. An algorithm attributes a relevance score to each report. Reports with a high relevance score are assigned a category while reports with a low relevance score are considered "trash". Reports with a relevance score between both thresholds are reviewed and classified through human analysis. To aid in this task, non-English articles are machine translated to English. This system then makes the categorized reports available to users in a variety of languages through machine translation.

Using interoperable services which are orchestrated as workflows, MEDCollector enables the flexible configuration of epidemic data collection from multiple sources. Collected data can then be packed into datasets for later use by epidemic modelling tools. Through the use of Web standards for data transmission the system can be seamlessly integrated with external Web services to extend its basic functionality.

MEDCollector not only gathers, but also integrates data from multiple and heterogeneous sources providing epidemiologists with not only the traditional data sources, but also a wide array of datasets obtained from the Web using its own data services.

3 Epidemic Data Collector Requirements

An epidemiological data collector should follow a set of principles and requirements that enable extensible data collection and the creation of consistent, integrated datasets, while coping with the heterogeneity associated with its sources.

These requirements can be functional, and therefore related to the how the system is used, or non-functional, which relate to the quality of the system.

3.1 Functional Requirements

Active Data Collection. Several Web sources, such as Twitter, currently expose a variety of APIs and URIs. Through the creation of mechanisms to access these APIs it is possible to actively collect their data. This enables data collection from sources like Twitter, Google Flu Trends and EISN reports. Depending on the source, the harvesting mechanism collects an entire message containing the name of a disease for further processing or harvest epidemiological estimates known to be published at the defined source.

Passive Data Collection. While using APIs specifically designed for data extraction is an efficient way to harvest data from several sources, some sources do not provide them and active data collection also creates a number of issues related to the number of requests to external services to keep data updated, such as limitations to the query rate and others. An approach to these problems is the exposure of passive collection interfaces which do not actively search for new data but wait for data to be submitted directly to them. One such example is an e-mail address, monitored by an application, through which the system can receive news and disease alert e-mail subscriptions (e.g. ProMED-mail). Data received by passive data collection mechanisms requires structuring before being integrated and loaded to the system.

Local Storage. Different data sources have variable data availability times, and data may only be available for some time period at certain sources, if any. An approach to solve the problem associated with dealing with volatile data, as well as the temporal disparity of data sources, is to locally store all the retrieved data in a local dedicated relational database.

Ontology Referencing. Enables the use of controlled vocabularies when referencing entities in the spatial and health domains. The use of ontologies enables the disambiguation of named entities, the mapping of entities with multiple references across data sources, and the establishment of hierarchical relationships between entities. This hierarchy becomes particularly relevant when using geographic referencing. For instance, with the support of a geographic ontology, we can relate cities with their respective countries. This enables the aggregation of data defined for specific levels to higher levels, e.g. disease cases identified in Lisbon can be used in Portugal's domain.

Use of Web Standards. Data transmission on the web requires the use of Web Standards, such as XML, enabling machines to parse its contents and process it accordingly. This enables the use of transmitted data independently of the devices or technologies used to access them.

Data Provenance. So that data can be validated for use in epidemic studies the epidemiologist must not only be able to know the source of the collected data but also be able to see and evaluate the mechanisms, or workflows, which have harvested the data. This way a user can restrict the data he uses in his research according to the harvesting mechanisms he knows to provide the most suitable data for that purpose.

Scheduling of Collection Activities. One of the challenges in data collection is the limit on the number of queries that can be performed during a period of time, which is imposed by the Web Source. Coping with this problem should require the data collection system to be able to define when data collection mechanisms are active on their sources and prioritize which queries are being performed at a given time. This improves the system's performance by improving the frequency of queries that provide the greatest amount of data while reducing the frequency of queries that provide little to no data.

Data Packaging. An epidemic data collection system is only useful to the epidemiologist community if it can be used to provide data for epidemic modelling and forecasting tools. For this purpose the data collection system should enable its users to query the collected data and create consistently packaged datasets.

3.2 Non-functional Requirements

Modularity and Configurability. An epidemic data collector that retrieves data from the Web requires a degree of flexibility in order to cope with changes or additions to its sources.

By adopting a Service-Oriented Architecture (SOA) [17], the system has its functionality distributed through discrete units, or services. SOA is implemented on the Web as HTTP-based RESTful Services or based on SOAP and WSDL which are considerably more complex. The use of standards permits unequivocal transmission of information by structuring the data clearly.

Workflows, or orchestrations as they are called in SOA, enable the design of data flow sequences between the different services. Configurable workflows enable the reconfiguration and addition of new services whenever necessary by defining how services are interconnected and how information is transmitted between them [18].

The Business Process Execution Language (BPEL) [19] is a XML-based language that describes the interactions between services, therefore allowing the definition of workflows. Each BPEL process is itself a web service. The BPEL process is described in XML and can be interpreted and executed by a BPEL engine. WSBPEL 2.0 is an OASIS standard that enables the definition of BPEL processes for web services under a SOA architecture.

Interoperability. Two types of interoperability are required. Firstly, interoperability between services within the system. By defining services with a set of configurable inputs and outputs, based on Web Standards, a high level of interoperability is possible, improving the flexibility of workflow creation. Secondly, interoperability between the system and other applications, such as epidemic forecasting tools. By enabling this type of interoperability, the system becomes more useful from an e-science perspective by enabling knowledge sharing between different systems.

Performance. Epidemic surveillance systems need to retrieve data from a large number of sources. Furthermore epidemic data source and news systems provide data more often, sometimes daily, and social networks provide it near "real-time". An epidemic data collection system should be able to cope with this large amount of data availability.

Fault Tolerance. Web Services can often fail due to a number of reasons such as time-outs (due to large numbers of accesses) or request limits. To make a workflow reliable in light of these problems it should be able to treat and respond to faults, e.g. treating a service fault resulting from rate-limiting measures and waiting the required amount of time to access the Web service again.

Data transmission on the web requires the use of Web Standards, such as XML, enabling machines to parse its contents and process it accordingly. This enables the use of transmitted data independently of the devices or technologies used to access them.

4 Architecture and Implementation

This section describes the design principles behind MEDCollector and its software implementation. MEDCollector implements the requirements detailed in Section 3. It is inspired in the "Data Collector" [20], an initial prototype for the collection of messages from Twitter containing disease and location names.

The Architecture of the MEDCollector is represented in Figure 4.

The main system components are:

- *Dashboard*. A web-based front end that provides user-interface capabilities to the system, enabling the user to define workflows for data harvesting and packaging.

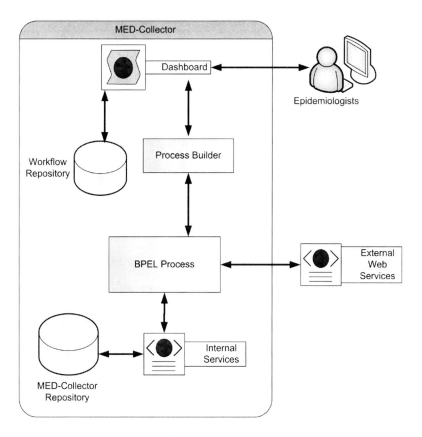

Fig. 4. MEDCollector's basic architecture

- *Workflow Repository.* Stores workflows designed through the Dashboard.
- *Process Builder.* Converts designed workflows to BPEL processes describing SOA orchestrations.
- *BPEL Processes.* Processes that are run on a BPEL engine and orchestrate communications between multiple services, both internal and external, to perform data collection or data packaging accordingly to workflow definition.
- *Internal Services.* Provide basic system functionalities and interact with the MEDCollector Repository.
- *External Services.* Available on the web, potentially extend the functionality of MEDCollector.
- *MEDCollector Repository.* Stores all the data collected by the system.

Users, design data collection mechanisms which are stored in a Workflow Repository. These mechanisms are processed into BPEL processes which orchestrate communication between services, both internal and external, performing epidemic data collection. All data collected by these Processes is stored on the MEDCollector Repository through an appropriate service.

This section is further organized in the following sub sections:

- 4.1 *Data Storage* - Describes the logical structure of the MEDCollector Repository.
- 4.2 *Ontologies* - Explains the ontologies used to reference entities in MEDCollector and how they are used.
- 4.3 *Dashboard* - Explores the implementation of the graphical user interface used for the design of workflows.
- 4.4 *BPEL Processes* - Describes de creation of a BPEL process from the initial design of the graphical workflow.
- 4.5 *Services* - Describes the several services currently implemented in MEDCollector.

4.1 Data Storage

All the data collected by the Multisource Epidemic Data Collector is stored in the MEDCollector repository to implement the requirement of local storage identified in Section 3. The choice of data sources is reflected in the structure and principles of Data Storage. MEDCollector's target sources differ greatly in the structure of their data as well the information contained in them. It focuses on extracting disease occurrences, incidences, estimates and messages in their geographical context. Since this data differs across sources, its storage's structure must be able to capture this heterogeneity.

An initial prospection of data sources revealed three main groups of possible epidemiological data sources:

- *Social Network Services*, such as Twitter, where people freely share information about themselves including their illnesses.
- *Epidemiological Surveillance Services*, such as EISN, ProMED-Mail, etc., which present very heterogeneous data. Each source has different data structures and formats.
- *New Services*, such as Google News, which report RSS feeds and newsletters containing news relating to specific domains and locations.

The MEDCollector Repository stores both the data collected from the Web and data collection schedules. It is implemented as a MySQL relational database. For clarity in the description of this repository's implementation, we present it as storage for two types of data: Case Data and Scheduling Data.

Case Data

The collected data is stored in the repository in a relational database. The central entity is a Case and it is related to an entry in each of the following entities:

- *Disease*, containing names of diseases and a concept unique identifier (CUI) that identifies that disease in the Unified Medical Language System (UMLS) [21].

– *Location*, containing data about the locations monitored by the system including a *geonameid* which identifies that location in the GeoNames ontology [22].
– *Source*, containing the sources used in MEDCollector, referenced by a URL and in some cases the update period of that source.
– *Process*, containing the data about the workflow which has generated it. This is relevant for data provenance analysis so that epidemiologists can evaluate which workflows provide the best data for their specific uses.

Besides these entities the compound key for a Case also includes the attributes *Date* and *Author*.

The attributes *value*, *unit* and *type* specify the type of case that was extracted, the value of that case and the unit used to express it e.g. In Twitter the system extracts 1 message, in Google Flu Trends the system can extract disease case estimates such as: 3 estimated cases per 100,000 population.

Scheduling Data

The schedule of data harvesting queries has an identical organization to the case data with the harvesting events as the central entity and related to the same entities, with the exception of the Process entity (Fig. 5(b)).

Scheduling Data is used by a priority based query selection service (See sub-section 4.5) to define which queries should be executed first at what sources.

Each week a background application updates the period values of each disease-location-source triple accordingly to the previous case entries in the last month:

– Schedule with a Daily period: every triple with more than 1 entry the previous week.
– Schedule with a Weekly period: every triple with more than 1 entry the previous two weeks and 1 or less entries the previous week.
– Schedule with a Fortnightly period: every triple with more than 1 entry the previous month and 1 or less entries the previous two weeks.
– Schedule with a Monthly period: every triple that does not fit the criteria mentioned above.

The disease, location and source entities in the repository are accessible through a series of services, these entities can be selected or inserted. The database currently includes all countries in the world and their capitals as well as a set of 89 infectious diseases.

4.2 Ontologies

The spatial and health-domain entities identified in the collected data are required to be referenced to a controlled and non-ambiguous vocabulary in order to create a common unequivocal language for all the system users. Ontologies are ideal for this type of problem since they provide a community reference,

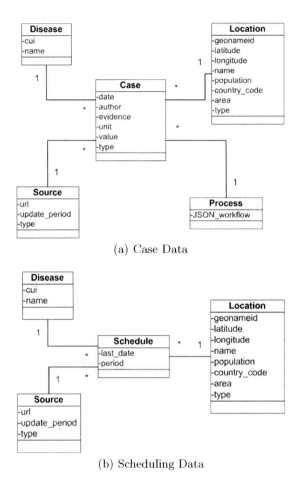

Fig. 5. UML class diagram of the MEDCollector Repository

improving knowledge reuse. This provides a common vocabulary through the use of non-ambiguous references, allowing all the users to use the same terms in relation to the same entities, and the establishment of hierarchical relations between entities. This hierarchy becomes particularly relevant when using geographic referencing. For instance, with the support of a geographic ontology, we can relate cities with their countries. This would allow data collected at a particular hierarchical level to be used at superior levels.

Therefore, cases identified by MEDCollector are referenced in the spatial and health domain through the use of ontologies.

Unified Medical Language System

The Unified Medical Language System (UMLS) is a collection of controlled vocabularies in the medical sciences which provides a structure for vocabulary mapping, allowing cross-translation of terms between terminologies [21]. It can also be seen as a multipurpose ontology of biomedical terms. Its purpose is to be used in biomedical informatics applications.

UMLS is composed of three Knowledge sources: the Metathesaurus, the Semantic Network and the SPECIALIST Lexicon.

The Metathesaurus is a vocabulary database that includes information about biomedical concepts, names and relationships. It is multi-lingual and is built from several terminologies, such as thesauri, classifications, code sets, lists of controlled terms, biomedical literature etc.

The Semantic Network provides categorization of all concepts represented in the Metathesaurus and the SPECIALIST Lexicon is a tool to enhance natural language processing.

Disease entities in the MEDCollector Repository are referenced to UMLS through a concept unique identifier (CUI) of a Metathesaurus concept categorized as a "Disease or Syndrome" in the Semantic Network.

GeoNames

Geonames is a database containing over eight million geographical names corresponding to over 6.5 million unique features [22]. Each of these features is categorized into one of nine classes and one of 645 feature codes. Geonames also includes latitude, longitude, elevation, population, administrative subdivision and postal codes for each of the features represented. Coordinates are represented according to the World Geodetic System 1984 [23].

Each GeoNames feature is represented via a GeonameID and a stable URI which gives access to an RDF description of the feature. The GeoNames Ontology describes feature properties using the Web Ontology Language (OWL). Feature classes and codes are described according to the Simple Knowledge Organization System (SKOS) [24]. GeoNames also provides a RESTful API to access feature properties and relations.

Each location in MEDCollector is referenced to a GeonameId. The MEDCollector currently includes all countries and capital cities in the world.

4.3 Dashboard

The collection of epidemic data requires a system with flexibility and configurability. A recognized approach to this problems is the use of scientific workflows which provide a declarative way of specifying complex mechanisms, or tasks, such as epidemic data collection mechanisms. Simple tasks can be implemented as local services and Web services, being responsible for small fragments of functionality. Scientific workflows enable the use of many such simple components to create complex mechanisms by chaining several such services in a specific order. This way, a data collection mechanism may be modified by the addition of new services or simply by adjusting parameters on the currently existing services, enabling the system to cope with the the addition of new data sources or to add further ways to process the collected data.

The Business Process Execution Language (BPEL) [19] is a workflow design language that uses XML to describe the interaction between services. The BPEL process, corresponding to a designed workflow, is itself a service.

One of the difficulties with the use of BPEL lies on the need of methods for creating process definitions by non-technical users [18]. Although graphical notations exist, such as the Business Process Modeling Notation (BPMN) [25], they are complex and require business analysts to design workflows. To overcome this issue MEDCollector has a Drag-and-Drop user interface, which enables users to design workflows by trial and error (Figure 6).

Scientific workflow systems like Taverna were considered, but currently they require the users to provide the WSDLs definitions of services and many of these systems also restrict communication protocols to SOAP, not allowing RESTful services. In addition, these systems currently do not offer on-browser Web interfaces, requiring users to go through lengthy installation and configuration processes prior to using the software.

WireIt[26] enables the definition of a "Visual Language" that specifies modules, their inputs and outputs, which represent services in MEDCollector. WireIt is an open source JavaScript library for the creation of web wirable interfaces similar to Yahoo! Pipes [27] and uDesign [28]. WireIt uses Yahoo! User Interface library 2.7.0 [29] for DOM [30] and event manipulation and is compatible with most web browsers. It is also bundled with a single-page editor that enables the definition of workflows through a wirable interface.

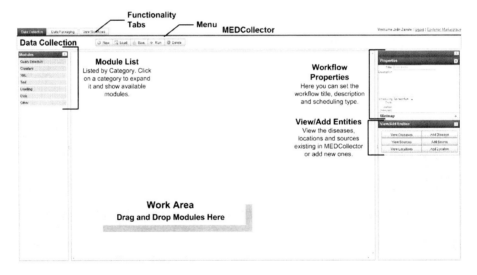

Fig. 6. Global view of the Web Interface implemented using WiringEditor and description of its components

Using the WireIt library, and based on the bundled single-page editor, a user interface was developed for MEDCollector.

This interface is composed of four sections (Figure 6):

- a menu which enables users to save, load and delete workflows.
- a properties section that enables users to configure the scheduling parameters for a workflow, its name and description.
- a modules section which enables the user to select which services to use in a workflow.
- a workspace area where the user defines a workflow by defining connections between service endpoints, by dragging graphical "wires" between their inputs and outputs.
- an entities section that can be used to browse the disease, location and source entities in MEDCollector and insert new ones.

The Dashboard has three tabs. These tabs are responsible for different functionalities in the system. The first is intended for data collection from the web and storage into a local database, while the second is designed for data packaging, enabling the user to query stored data into downloadable datasets. A third tab which is composed only of a loading menu and a work area enables users to view workflows designed by other users while not enabling them to perform alterations. This enables users to select which workflows to extract information from, to check the provenance of data by analysing the workflows that originated it, or simply to gather design ideas for their own workflows.

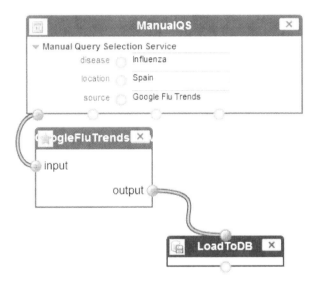

Fig. 7. A simple workflow designed in MEDCollector for the collection of data from Google Flu Trends

4.4 BPEL Processes

Worklows designed in the Dashboard are stores in the workflow repository as JavaScript Object Notation (JSON) documents containing the properties of the workflow, a list of used services, or modules, and a list of wirings between these services.

Figure 7 presents an example of a workflow designed in MEDCollector's Dashboard, it is stored as the following JSON document:

```
{
    "modules":[
      {
        "config":{"position":[113,63]},
        "name":"ManualQS",
        "value"  :{
          "disease":"Influenza",
          "location":"Spain",
          "source":"Gloogle Flu Trends"
        }
      },
      {
        "config":{
          "position":[85,286],
          "xtype":"WireIt.InOutContainer"},
        "name":"GoogleFluTrendsHarvest",
        "value":{}
      },
      {
        "config":{
          "position":[269,295],
          "xtype":"WireIt.InOutContainer"},
        "name":"LoadToDB",
        "value":{}
      }
    ],
    "properties":{
      "description":"",
      "name":"WorkflowName",
      "owner":"Username",
      "schedule-time":"",
      "scheduling-type":"Do Not Run"
    },
    "wires":[
      {
        "src":{"moduleId":0,"terminal":"XMLOutput"},
        "tgt":{"moduleId":1,"terminal":"input"}
      },
```

```
      {
          "src":{"moduleId":1,"terminal":"output"},
          "tgt":{"moduleId":2,"terminal":"input"}
      }
   ]
}
```

These JSON documents are parsed by the Process Builder, which interprets this document to create the files necessary to deploy a BPEL Process. This process consists of a process descriptor, a XML BPEL process definition and a WSDL document describing it as a service. In the BPEL process, each service is represented as a PartnerLink and the wirings are mapped as a sequence of variable assignments and PartnerLink invocations.

BPEL Processes, executed by a BPEL Engine, orchestrate communications between services, to perform data collection accordingly to the workflow definition.

These BPEL process can have two functions, data collection or data packaging, according to the dashboard tab they are designed in.

A series of Fault handling mechanisms are also added to the BPEL processes to recover from failed invocation of services which can happen for several reasons, including invocation rate limiting measures implemented by several sources. For instance when RESTful Web services return an HTTP code 503, the status code for service unavailable, the process stops and delays its next scheduled execution.

Apache ODE (Orchestration Director Engine) [31] is used to execute the BPEL Processes. Apache ODE provides several extensions to standard BPEL engines including XPath 2.0 support, for easier variable assignments, and an HTTP binding extension that enables direct connection to RESTful Web Services. This engine also provides an interface that enables monitorization of currently deployed processes.

4.5 Services

Services represent fragments of functionality, or simple tasks which can be performed by the workflow system. These can be information collection services, text mining services, transformation services, scheduler services, and others. All services are currently implemented in PHP Version 5.1.6.

Implemented services are divided into three categories: Data Collection Services, Data Packaging Services and Generic Services.

Data Collection Services

These services are used for the design of data collection workflows. They include several tasks, from the selection of which entities to search for, to retrieving the data as well as data transformation and loading data to the MEDCollecto Repository.

Query Selection Services are services that choose the query (disease, location and source triples) to be made by other services. There are two types of query selection services:

- Priority Based Query Selection: This service uses the schedule data to prioritize the queries. It selects the period and last search date values, from the Scheduling Data in MEDCollector's repository, for each disease-location-source triple and outputs the triple with the highest positive priority value according to the formula:

$$priority = date - last\ search\ date - period$$

  If there are no positive values the service sends a fault message that is caught by the BPEL Engine, stopping it and scheduling another run of the service later. These triples can be filtered by source, location or disease, in order to create processes with specific scopes, e.g. influenza in Portugal with Twitter as a source.
- Manual Query Selection: This service enables users to manually define the disease and location to query at a source.

Data Harvesting Services are services specifically designed to gather data through specific APIs or URLs. The Query Selection Services define what queries these harvesting services perform at their sources.

Currently there are two harvesting services implemented, one to monitor Twitter and one that harvest data from Google Flu Trends.

- TwitterHarvest: This service receives a a message from a Query Selection service specifying which location and disease to search for in Twitter. The service uses the Twitter search API to retrieve messages containing the disease name and posted at a the location specified by the Query Selection service.
- GoogleFluTrendsHarvest: This service receives a message from the Query Selection service specifying which locations to search for Influenza data. Using the country-code for that location it retrieves a text file containing comma separated values (Through the URL http://www.google.org/flutrends/<country-code>/data.txt). It retrieves the values of estimated influenza activity from the column containing the location name and its respective date.

These services structure and transmit the collected cases using a XML schema compatible with other MEDCollector services.

Passive Collection Services receive data posted by disease reporting services and other sources such as e-mail subscriptions. Currently there is a service that checks an e-mail for messages containing names of diseases and locations. This service checks for messages that have not been read and searches them for all locations and diseases in the MEDCollector Repository. If a match is found the message is retrieved and referenced to that location and disease. This service also structures and transmits the collected cases using a XML schema compatible with other MEDCollector services.

Text related services include a regular expression matching service that searches strings for patterns, text mining service and translation services:

- Text Mining services receive a message from harvesting and passive collection services and through a rule based approach tries to match sentences of these messages to a number of regular expressions. These regular expressions enable the extraction of case numbers, estimates and deaths. Although this rule-based approach and some of the regular expressions might not be the best approach for Text Mining in this field, this service serves as proof-of-concept for the extraction of further information from text messages in a workflow oriented data collection system. The development of an appropriate Text Mining algorithm is beyond the scope of this paper.
- Translation services use the REST services in the Google Language API [32]. There are two services, one that translates a string and another that given a XML element path translates the text node inside that element. Both these services require the user to specify a desired output language in the two letter code specified in BCP57 [33]. The use may also specify the language of the original message, however if this parameter is not provided the service will use the Google Language API to try to identify the message's original language.

Database Loading is done through a service that receives a XML message and accordingly performs an insertion in the MEDCollector Repository. Upon insertion the service will return the XML document back to the caller.

XML Transformation Services which enable the transformation of XML documents into a schema compatible with other MEDCollector Services. This enables the use of external web services in the sequence flow by transforming their data into compatible data types.

There are several services for this task, one that transforms RSS 2.01 feeds and another that transforms Atom 1.0 feeds. These receive the source name and optionally a disease and location. If that disease and location is given, then all the entries in the feed are considered cases of messages pertaining to that disease-location-source triple, otherwise each entry is searched for all disease-location pairs to make that correspondence. Disease and location entities are searched using a binary search algorithm. Another XML Transformation Service that is more complex, and less user-friendly, requires that the user provide the paths to the relevant parts of the document, enabling the transformation of all other XML documents available. This is done using XPath notation.

Data Packaging Services

Data Packaging services offer the functionality to be used in workflows for dataset creation as well as dataset upload to EPIWORK's information platform, the Epidemic Marketplace.

Data Query Service. This is a service that has a number of optional parameters enabling the user to query the MEDCollector Repository for the data required for dataset creation.

Data Formatting Services. By default the queried data is transmitted in the MEDCollector XML schema. Through the use of these services it is possible to convert that XML into JSON and CSV documents. The JSON document keeps the overall hierarchy of the XML document while the CSV transformation service transforms it into a readable two-dimensional matrix. The CSV transformation service also enables the user to specify which delimiters to use for rows and columns.

Download Service. This simple service that enables the user to download the resulting document instead of uploading it to the Epidemic Marketplace.

EM Upload Service. A service which enables users to upload datasets to the Epidemic Marketplace repository instead of simply downloading them. This service has several parameters which can be filled out as metadata for the dataset.

Generic Services

Generic Services are services that can be used in both the Data Collection and Data Packaging workflow design and are therefore available on both tabs of the Dashboard.

Date Related Services are services that relate to time formats and current dates. The "Date Format" service provides the user with the capability of transforming any date to a specific format. MEDCollector uses "Y-m-d" by default so this service enables users to create data collection mechanisms that transform the date so it can be used by MEDCollector, or to transform the date in the extracted datasets into the formats used by their applications. The other Date service returns the current date in the provided format ("Y-m-d" by default). This enables the user to specify provide services with the time at which the process is run. For instance if a XML document does not have a time reference this service can be used to provide that reference, likewise this service can be used to provide the Data Querying services with the current time, so that it retrieves data in a period of time related to the date of each run. Formats are specified in PHP notation[1].

XML Manipulation Services provide functionalities such as data structure transformation and access to specific data elements to improve the flexibility of the workflows. These include:

---

[1] http://php.net/manual/en/function.strftime.php

- A *GetElement* service, which returns the node specified by a XPath query from a given input XML document.
- A *Merge* and *Split* gates. The first is given two XML document as inputs, it merges them at the Root Level. The second splits a XML document into two by evaluating condition on a node from each entry. This condition can be of the types: "equals", "contains", ">=", "<=", ">" or "<". The Split gate returns a XML document for the entries evaluated as true and another for the entries evaluated as false.

External Services can be used to provide additional functionalities to the system. A SOAP and a REST wrapper services are available for this purpose.

The SOAP service receives a WSDL URL, an operation name and a set of parameter names and values, which correspond to WSDL message parts. This service sends a SOAP call and returns the response to that call.

The REST service receives a URL, a query string and a set of parameter names and values. These are then sent through cURL [34], a command line tool data transfer with URL syntax, and the response is then returned. If the cURL request returns an HTTP code other than 200 it generates a Fault corresponding to that HTTP code to be interpreted by the BPEL Process.

5 Evaluation and Discussion

This section evaluates the system according to the functional and non-functional requirements set on section 3. For this purpose *MEDCollector* is compared with two known scientific workflow design programs, *Taverna* [9] and *Kepler* [11].

MEDCollector is presented as a workflow design environment with a set of predefined services that provide its basic functionalities according to the requirements previously mentioned.

Taverna, does not provide any services itself, instead it provides the software to access WSDL-based web services. For this reason, *Taverna* used the services developed for *MEDCollector* and because of this it presents the same basic functionalities in this evaluation. For this reason Taverna is only compared to MEDCollector on the functional analysis in regard to data provenance.

Kepler is a scientific workflow design tool for data intensive processes and presents a large variety of actors that provide data flow functionalities. *Kepler* is compared with *MEDCollector* on both the functional and non-functional analysis.

For this comparison data harvesting workflows were designed to harvest cases from:

1. *CDC Flu Updates RSS Feed* - A workflow retrieves the feed and inserts the items into the database as messages about Influenza in the United States.
2. *CDC Travel Notices RSS Feed* - A workflow retrieves the feed, searches the items for disease and location entities and if these are found the item is inserted into the database.

3. *Google Flu Trends* - A workflow retrieves Influenza estimates and inserts them into the database. Spain was used as the location for the data collection in this workflow.
4. *E-mail Messages* - A workflow retrieves a message from the e-mail server, searches it for disease and location entities and if they are found it inserts the message into the database.

5.1 Functional Analysis

In this section we compare the functional aspects of *MEDCollector* with the scientific workflow design software *Kepler*. Table 1 presents an overview of the functional analysis. Kepler provides a vast array of functions and services, however, since the scope of this paper is epidemic data collection this analysis will only focus on workflows designed for that purpose.

Table 1. Functional Analysis overview. "Yes" means the functional requirement is fulfilled, and "No" means it is not fulfilled. "Yes*" means the functional requirement is fulfilled but requires more technical knowledge than MEDCollector.

System	Active D.C.	Passive D.C.	Local Storage	Web Standards	Data Provenance	Scheduling	Data Packaging
MEDCollector	Yes	Yes	Yes	Yes	Yes	Yes	Yes
Kepler	Yes	No	Yes*	Yes*	No	Yes	Yes*
Taverna	-	-	-	-	Yes	-	-

Active Data Collection - MEDCollector provides active data collection functionality through a number of services for specific web sources or by enabling users to use REST or WSDL Web services as data collection sources. Likewise, Kepler enables the user to use service outputs as input of data for its workflows through a series of actors that enable the user to access data from a variety of sources, such as RESTful and WSDL web services.

Passive Data Collection - MEDCollector enables passive data collection by enabling the extraction of e-mail messages and by enabling users to add collector records through services specifically designed for this function. Kepler, however, does not provide passive data collection actors that enable the creation of workflows for e-mail message collection or other types of passive data collection.

Local Storage - Data is stored locally, on the server in a relational database. In MEDCollector, this is done through a database loading service, specifically designed for this purpose which connects to the database and inserts the data.

Kepler also enables the user to connect and query a relational database. However, Kepler requires the user to define the connection to the database, therefore the user needs to have knowledge of specific details of the database server such as its address and user credentials for the database. In Kepler the user also needs to compose the query to be performed on the database, requiring the user to have technical knowledge of SQL (Figure 8). Furthermore, for Kepler to be able to access a database it needs to have specific endpoints for the Java Database Connectivity (JDBC) driver - e.g. MySQL is required to have the Connector/J plugin.

In MEDCollector the user is not required to have technical knowledge of the database server as all operations are performed by the back end web services.

Use of Web Standards - The use of Web Standards enables system interoperability and data gathering from multiple sources.

Both MEDCollector and Kepler are able to use Web Standards, however their implementations are significantly different. While MEDCollector uses dedicated services for the transformation of data received from its sources, Kepler is able to access XML document elements through workflows that the user designs to access those elements, e.g. by assembling and disassembling XML documents and through arrays returned by XPath query actors (see Fig. 8).

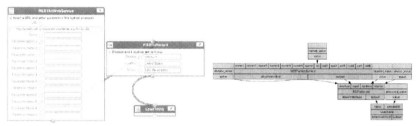

(a) CDC Flu Updates Workflow in MEDCollector.

(b) CDC Flu Updates Workflow in Taverna.

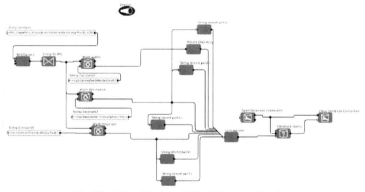

(c) CDC Flu Updates Workflow in Kepler.

Fig. 8. Workflows designed for the extraction of messages from CDC Flu Updates RSS feed. This feed contains only messages relating to Influenza in the United States.

The issue with this second approach is that it leads to more complex workflows, with more actors, shifting the user's focus from collecting the data to transforming it.

After data transformation, MEDCollector services use the same XML schema and character strings for inputs and outputs. Kepler is more complex having several different data types, e.g. arrays, booleans, strings, XML Elements, etc. For this reason the user needs to convert data types to create the workflows he wants to design depending on the inputs and outputs of each actor.

Data Provenance - Each workflow in MEDCollector is identified by a process id. This id is then used to reference each case collected by that workflow in the MEDCollector Repository. Using the *View Workflows* tab on the Dashboard, users can see the workflows that were used to collect data. Also, logging files are kept on Apache ODE at all times so that it is possible to pinpoint any issues in a specific workflow or a faulty service.

Taverna provides a detailed view of the workflow's execution in its *"Results"* tab. This tab enables the user to see the inputs and outputs of each service as well as its execution times.

Currently, Kepler's provenance framework is still under development and its functionalities are not currently accessible to the general user.

Scheduling of Collection Activities - MEDCollector provides this function based on two approaches, scheduling of processes and prioritization of queries. The first enables users to define when a service is actively collecting data from its source with its period of activity specified in minutes. The latter enables the workflows to use queries that provide most messages to be harvested more often than queries that provide no messages.

Kepler uses two types of entities, actors and directors. Actors perform specific functionalities while directors coordinate when actor's functions are performed. The different types of directors define if workflows are run synchronously or asynchronously, the number of iterations, and when they are executed.

Data Packaging - MEDCollector provides a Data Packaging tab on the Dashboard which provides a number of services that enable the extraction of detected cases from the MEDCollector Repository. Accessing the local repository is done through a Query service, to which the user provides a number of optional parameters for data filtering and transformation is also provided through a set of services.

In Kepler the user is required to write the Query in SQL and designed the workflow to deal with all the data transformation as described earlier. This is a complex task which requires the user to have technical knowledge of the database implementation as well as to deal with multiple data types and their transformation.

5.2 Non-functional Analysis

Here we evaluate MEDCollector from a non-functional perspective, comparing it to Kepler and Taverna. We focus on the non-functional aspects that can be better compared between these systems, such as performance, modularity and configurability.

Performance

All the workflows designed for this evaluation were executed separately. All workflows were executed in the same machine, with a 2Ghz CPU and 1 Gigabyte of RAM, on the CentOS 5.5 distribution of the Linux Operating System. MEDCollector Services were hosted on the Apache 2.2.3 Web Server. Apache ODE 1.3.3 was hosted on Apache Tomcat 6.0. Databases were hosted on MySQL Server 5. The server also included PHP 5.1.6 and Java 1.6.0_21.

The results in Figures 9 and 10 are average run times over 5 executions for each workflow.

An initial analysis of the execution times shows an apparent disparity in Kepler's execution times in relation to those of MEDCollector and Taverna throughout the different workflows. This disparity is explored below with some insights into the designed workflows and their executions.

CDC Flu Updates RSS Feed Workflow and Google Flu Trends Workflow

In these two workflows Kepler performed best with only a fraction of the execution times from MEDCollector and Taverna. MEDCollector was slower performing closely to Taverna.

While MEDCollector's and Taverna's functionalities are provided by services that require communication through a Web server, Kepler's actors are Java classes that are directly invoked in its engine. Evidence from these two workflows shows that this direct invocation results in lower latency between the invocation of actors and therefore improves the overall performance of workflows.

CDC Travel Notices RSS Feed Workflow

In this workflow the Kepler performed worst with MEDCollector and Taverna performing similarly. CDC's Travel Notices can have several distinct subjects and as such messages have to be searched for evidence of locations and diseases. For this purpose messages are searched for disease and location names in the MEDCollector repository.

Both MEDCollector and Taverna use an RSS transformation service which, when not provided with a disease and location name, uses binary search algorithm to search for those entities.

Kepler's actors do not allow the replication of this algorithm, and as such this workflow is required to iterate over all disease and location pairs for each message. In this workflow we have 89 diseases and 379 locations which results in 33731 disease-location pairs that need to be searched for each message. This results in an exponentially larger workflow run time as messages, disease or locations increase, which explains the dramatically larger execution time in the kepler workflow.

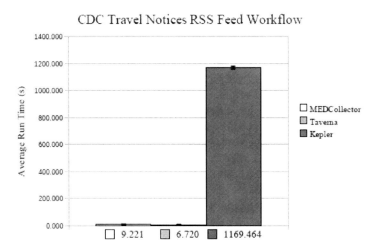

(a) Bar chart representing average workflow run times for CDC Travel Notices RSS Feed.

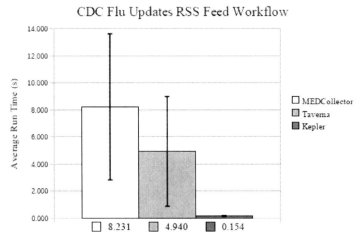

(b) Bar chart representing average workflow run times for CDC Flu Updates RSS Feed.

Fig. 9. Performance charts of the CDC Travel Notices RSS Feed and CDC Flu Updates RSS Feed workflows. Average run times are presented in seconds rounded to three decimal places.

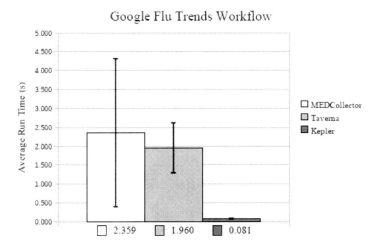

(a) Bar chart representing average workflow run times for Google Flu Trends.

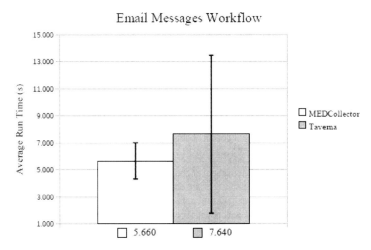

(b) Bar chart representing average workflow run times for e-mail Message retrieval.

Fig. 10. Performance charts of the Google Flu Trends and e-mail message workflows. Average run times are presented in seconds rounded to three decimal places.

E-mail Messages Workflow

While MEDCollector and Taverna use services developed to extract messages from an e-mail server, kepler does not provide such functionality. Due to this, Kepler is excluded from this analysis. Taverna and MEDCollector did not perform significantly different.

Overall, Taverna and MEDCollector had similar performances throughout the workflows. The reason for this lies in that they both use the same services, and both invoke them through a web server. For this reason the latency between service invocation is very similar. MEDCollector's workflow execution engine is Apache ODE which executes BPEL workflows and runs on Apache Tomcat. Taverna uses the Scufl Workflow Language [9] and its engine runs directly on the Java environment. This different back-end implementations explain the small differences in performance show above.

Kepler's actors are directly invoked as Java classes and functions which results in shorter latency times, improving the general performance of its workflows. However, MEDCollector's services were specifically designed for epidemic data collection and in some cases they perform better than Kepler. Further more Kepler does not provide the functionalities necessary for passive data collection, as seen in the *E-mail Messages Workflow*.

Modularity and Configurability

MEDCollector's modularity and configurability are provided by its services and the workflow's scheduling properties. Services were designed for epidemic data collection, performing the functionalities needed by users to achieve this task. This enables users to create workflows by focusing on the actions they need to perform to retrieve data from its sources. A service in MEDCollector can perform several actions on the data, e.g. the RSS transformation service in MEDCollector accesses different nodes in the XML tree, can search the description of each RSS item for a location and disease, and transform the relevant items into MEDCollector records in another XML schema.

Taverna's functionalities, as well as its modularity and configurability are dependent on the services it uses. Since it does not come packaged with a set of services that can be used for epidemic data collection the services developed for MEDCollector were used in Taverna, therefore providing it the same characteristics as MEDCollector in terms of modularity and configurability (see Figure 8 a and b).

Kepler is modular and configurable on a lower level of abstraction. Kepler's workflows are focused on the data, each actor is a single action on that data. To perform the same actions as a single MEDCollector service the user is required to use several actors in Kepler. As an example, to perform the RSS transformation the user would be required to use Xpath query actors to access the XML nodes, database connection and querying actors to retrieve a list of diseases and locations, string matching actors to search for those entities, and boolean control actors to separate the actors that matched the entities from those that did not. Kepler's actors specificity enables it to perform a broader number of functions in scientific workflows, however, this excess in modularity for the purpose of epidemic data collection results in much more complex workflows which shift the user's focus from data collection to single data transformation actions (see Figure 8).

5.3 Discussion

Overall all three systems had their strengths and faults. However, MEDCollector fulfilled all the requirements identified in Section 3. The Kepler scientific workflow system is highly focused on data operations and performs best at simple tasks due to its data-driven architecture. However, the task of epidemic data collection from multiple Web sources requires a high level of data treatment and transformation that requires the use of a large number of actors resulting in largely complex workflows. Furthermore, several of the actions require that users have knowledge of the technical implementations of servers, as is the case for database servers. Furthermore, Kepler does not provide actors that allow the definition of passive that collection workflows.

Taverna is a workflow design and execution environment. It is similar to MEDCollector in which its functionalities are provided by Web services. However, since Taverna is a general purpose workflow design software it does not provide a basic set of services. As a result Taverna requires users to specify directly what Web services are used through WSDL files.

MEDCollector takes a similar approach, however it provides users with the basic services for epidemic data collection. Some of these services are designed specifically to simplify the system's use and improve performance when searching for unknown disease or location entities. With MEDCollector's services the user can focus on the collection of the data instead of having to focus on the specificities of complex data transformation tasks as in Kepler. Furthermore MEDCollector's interface is Web based and as such does not require users to go through lengthy installations and to worry about specific configurations for their machines, such as the definition of proxies. MEDCollector is a workflow definition tool through a graphical interface that was specifically designed for epidemic data collection. It enables the flexible configuration of data collection mechanisms and is able to perform active and passive data collection.

6 Conclusions and Future Directions

The MEDCollector is implemented as a component for the EPIWORK project information platform - the Epidemic Marketplace. By enabling the collection and integration of data from multiple Web sources, MEDCollector grants epidemiologists with a novel means to gather data for use in epidemic modelling tools.

While manual curation, such as the one used by GPHIN, usually provides more accurate data, resulting from direct human analysis, it is expensive and limited in terms of the volume of data it can process. MEDCollector takes a different approach from GPHIN. Instead of manually classifying the data it requires users to design automated data collection workflows for each specific source, adapting the extraction mechanism to its nature therefore improving the automatic data extraction from each individual source. These collection workflows, retrieve all identified cases at their source and insert them into the MEDCollector Repository.

The foremost innovation brought by this system is that, through a browser based Dashboard, epidemiologists can design simple Web Service workflows dynamically using drag-and-drop components. This enables users to directly create and modify workflows to customize data collection mechanisms according to their specific needs. This dynamic design grants the system flexibility by enabling the addition of further functionality and the alteration of service interactions.

While MEDCollector can perform less efficiently than Kepler, it enables users with less technical knowledge of the system's implementation to design workflows for epidemic data collection using less workflow components. Furthermore MEDCollector's dashboard enables users to interactively browse the workflows developed by other users and extract the data collected by those workflows.

An interface layer that accommodates the configuration of dataset creation services enables the selection of information from MEDCollector's Repository and structure it according to the needs of the user, through XML transformation and querying. This transformation enables the creation of aggregated and consistent datasets which can be used by other applications.

Through the use of Web Standards for data transmission MEDCollector enables seamless integration of externally supplied Web services, granting extensibility to its basic features. This extensibility enables epidemiologists to design extraction mechanisms that suit their needs in extracting data from multiple Web sources.

Furthermore, EPIWORK's Epidemic Marketplace includes a data repository where datasets can be stored for later use by epidemic modelling tools. Mediator services were developed to enable applications to perform actions on the repository, such as uploading datasets. This mediator enables MEDCollector to submit packaged datasets for storage in the Epidemic Marketplace at regular time periods through the design of data packaging workflows, as opposed to simply downloading them.

The multisource epidemic data collector will therefore be a useful complementation to national reporting systems and hopefully a contribution to disease propagation studies as well as disease control research.

A set of challenges remain to be explored in future work. While MEDCollector enables the collection and packaging of data a set of visualization tools should be developed in order to improve the graphical information given to users upon the creation of datasets. This should provide users with the ability to quickly analyse the data so they can improve their datasets.

MEDCollector extracts data about putative infections and references them geographically, furthermore the author of extracted messages is also referenced. This leads to a number of privacy related issues, particularly considering MEDCollector was developed for data sharing across the Web. A number of anonymization techniques should be developed and implemented for epidemic datasets in general and MEDCollector in particular to protect the privacy of the individuals that data represents.

Acknowledgements. The authors want to thank the European Commission for the financial support of the EPIWORK project under the Seventh Framework Programme (Grant # 231807), the EPIWORK project partners, CMU-Portugal partnership and FCT (Portuguese research funding agency) for its LaSIGE Multi-annual support.

References

1. Brownstein, J., Freifeld, C.: HealthMap: The development of automated real-time internet surveillance for epidemic intelligence. Euro. Surveill. 12(10), 71129 (2007)
2. Ginsberg, J., Mohebbi, M., Patel, R., Brammer, L., Smolinski, M., Brilliant, L.: Detecting influenza epidemics using search engine query data. Nature 457(7232), 1012–1014 (2008)
3. Mawudeku, A., Blench, M.: Global Public Health Intelligence Network (GPHIN). In: 7th Conference of the Association for Machine Translation in the Americas, pp. 8–12 (2006)
4. Van Noort, S., Muehlen, M., Rebelo, A., Koppeschaar, C., Lima, L., Gomes, M.: Gripenet: an internet-based system to monitor influenza-like illness uniformly across Europe. Euro. Surveill. 12(7), 5 (2007)
5. Twitter, http://www.twitter.com/ (accessed June, 2011)
6. Silva, M.J., Silva, F.A., Lopes, L.F., Couto, F.M.: Building a digital library for epidemic modelling. In: Proceedings of ICDL 2010, The International Conference on Digital Libraries, February 23-27, vol. 1, TERI Press, New Delhi (2010)
7. Zamite, J., Silva, F.A.B., Couto, F., Silva, M.J.: MEDCollector: Multisource epidemic data collector. In: Khuri, S., Lhotská, L., Pisanti, N. (eds.) ITBAM 2010. LNCS, vol. 6266, pp. 16–30. Springer, Heidelberg (2010), http://dx.doi.org/10.1007/978-3-642-15020-3_2
8. Noronha, N., Campos, J.P., Gomes, D., Silva, M.J., Borbinha, J.L.: A deposit for digital collections. In: Constantopoulos, P., Sølvberg, I.T. (eds.) ECDL 2001. LNCS, vol. 2163, pp. 200–212. Springer, Heidelberg (2001), http://dx.doi.org/10.1007/3-540-44796-2_18
9. Li, P., Castrillo, J., Velarde, G., Wassink, I., Soiland-Reyes, S., Owen, S., Withers, D., Oinn, T., Pocock, M., Goble, C., Oliver, S., Kell, D.: Performing statistical analyses on quantitative data in taverna workflows: an example using r and maxdbrowse to identify differentially-expressed genes from microarray data. BMC Bioinformatics 9(334) (August 2008)
10. Gibson, A., Gamble, M., Wolstencroft, K., Oinn, T., Goble, C.: The data playground: An intuitive workflow specification environment. In: IEEE International Conference on e-Science and Grid Computing, pp. 59–68 (2007)
11. Ludäscher, B., Altintas, I., Berkley, C., Higgins, D., Jaeger, E., Jones, M., Lee, E., Tao, J., Zhao, Y.: Scientific workflow management and the Kepler system. Concurrency and Computation: Practice and Experience 18(10), 1039–1065 (2006)
12. Riedel, M., Memon, A., Memon, M., Mallmann, D., Streit, A., Wolf, F., Lippert, T., Venturi, V., Andreetto, P., Marzolla, M., Ferraro, A., Ghiselli, A., Hedman, F., Shah, Z.A., Salzemann, J., Da Costa, A., Breton, V., Kasam, V., Hofmann-Apitius, M., Snelling, D., van de Berghe, S., Li, V., Brewer, S., Dunlop, A., De Silva, N.: Improving e-Science with Interoperability of the e-Infrastructures EGEE and DEISA. In: International Convention on Information and Communication Technology, Electronics and Microelectronics (MIPRO), Opatija, Croatia, pp. 225–231 (2008)

13. Madoff, L., Yu, V.: ProMED-mail: an early warning system for emerging diseases. Clinical Infectious Diseases 39(2), 227–232 (2004)
14. European Center for Disease Prevention and Control (ECDC), http://www.ecdc.europa.eu/ (accessed June, 2011)
15. European Influenza Surveillance Network (EISN), http://www.ecdc.europa.eu/en/activities/surveillance/EISN/ (accessed June, 2011)
16. Marquet, R., Bartelds, A., van Noort, S., Koppeschaar, C., Paget, J., Schellevis, F., van der Zee, J.: Internet-based monitoring of influenza-like illness(ILI) in the general population of the Netherlands during the 2003 - 2004 influenza season. BMC Public Health 6(1), 242 (2006)
17. Durvasula, S., Guttmann, M., Kumar, A., Lamb, J., Mitchell, T., Oral, B., Pai, Y., Sedlack, T., Sharma, H., Sundaresan, S.: SOA Practitioners Guide, Part 2, SOA Reference Architecture (2006)
18. Garlan, D.: Using service-oriented architectures for socio-cultural analysis, http://acme.able.cs.cmu.edu/pubs/show.php?id=290
19. Alves, A., Arkin, A., Askary, S., Bloch, B., Curbera, F., Goland, Y., Kartha, N., Sterling, König, D., Mehta, V., Thatte, S., van der Rijn, D., Yendluri, P., Yiu, A.: Web services business process execution language version 2.0. OASIS Committee Draft (May 2006)
20. Lopes, L.F., Zamite, J., Tavares, B., Couto, F., Silva, F., Silva, M.J.: Automated social network epidemic data collector. In: INForum - Simpósio de Informática (September 2009)
21. Bodenreider, O.: The unified medical language system (umls): integrating biomedical terminology. Nucl. Acids Res. 32(suppl. 1), D267–D270 (2004), http://dx.doi.org/10.1093/nar/gkh061
22. GeoNames, http://www.geonames.org/ (accessed June, 2011)
23. Decker, B.: World geodetic system 1984 (1986)
24. Miles, A., Matthews, B., Wilson, M., Brickley, D.: SKOS Core: Simple knowledge organisation for the web. DCMI 5, 1–9
25. Business process modeling notation (bpmn) version 1.2, Tech. Rep. (January 2009), http://www.omg.org/spec/BPMN/1.2/PDF
26. Aboauf, E.: WireIt - a Javascript Wiring Library, http://javascript.neyric.com/wireit/ (accessed June, 2011)
27. Yahoo Pipes, http://pipes.yahoo.com/pipes (accessed June, 2011)
28. Sousa, J., Schmerl, B., Poladian, V., Brodsky, A.: uDesign: End-User Design Applied to Monitoring and Control Applications for Smart Spaces. In: Proceedings of the 2008 Working IFIP/IEEE Conference on Software Architecture (2008)
29. Yahoo User Interface Library, http://developer.yahoo.com/yui/ (accessed June, 2011)
30. Le Hors, A., Le Hégaret, P., Wood, L., Nicol, G., Robie, J., Champion, M., Byrne, S.: Document object model (DOM) level 3 core specification. W3C Recommendation (2004)
31. T. A. S. Foundation. Apache Orchestration Director Engine, http://ode.apache.org/ (accessed June, 2011)
32. Google AJAX Language API, http://code.google.com/apis/ajaxlanguage/ (accessed June, 2011)
33. Witt, A., Sasaki, F., Teich, E., Calzolari, N., Wittenburg, P.: Uses and usage of language resource-related standards. In: LREC 2008 Workshop (2008)
34. cURL, http://curl.haxx.se/ (accessed June, 2011)

Supporting BioMedical Information Retrieval: The BioTracer Approach[*]

Heri Ramampiaro[1] and Chen Li[2]

[1] Department of Computer and Information Science
Norwegian University of Science and Technology (NTNU)
N-7491, Trondheim, Norway
heri@idi.ntnu.no
[2] Dept. of Computer Science
University of California, Irvine (UCI)
Irvine, CA 92697-3425, USA
chenli@ics.uci.edu

Abstract. The large amount and diversity of available biomedical information has put a high demand on existing search systems. Such a tool should be able to not only retrieve the sought information, but also filter out irrelevant documents, while giving the relevant ones the highest ranking. Focusing on biomedical information, this work investigates how to improve the ability for a system to find and rank relevant documents. To achieve this goal, we apply a series of information retrieval techniques to search in biomedical information and combine them in an optimal manner. These techniques include extending and using well-established information retrieval (IR) similarity models such as the Vector Space Model (VSM) and BM25 and their underlying scoring schemes. The techniques also allow users to affect the ranking according to their view of relevance. The techniques have been implemented and tested in a proof-of-concept prototype called BioTracer, which extends a Java-based open source search engine library. The results from our experiments using the TREC 2004 Genomic Track collection are promising. Our investigation have also revealed that involving the user in the search process will indeed have positive effects on the ranking of search results, and that the approaches used in BioTracer can be used to meet the user's information needs.

Keywords: Biomedical Information Retrieval, Evaluation, BioTracer.

1 Background and Motivation

The continuous increase in the amount of available biomedical information has resulted in a higher demand on biomedical information retrieval (IR) systems. While their use has helped researchers in the field to stay updated on recent literature, many of the existing search systems tend to be either too restrictive (returning results with a low recall) or too broad (finding results with a

---

[*] This article is a revised and an extended version of the ITBAM 2010 paper [1].

low precision). For this reason, there is a need to improve existing search systems, especially with respect to retrieval performance, in order to improve their precision and recall.

1.1 Challenges for Biomedical Information Retrieval

From the information retrieval point of view, there are many challenges in retrieving biomedical information. First, as in most scientific domains, there is a wide use of domain-specific terminology [2]. For example, most of the documents extensively use specific gene names, names of diseases and/or specific biomedical specific terms as part of the text. Most methods should take this into account to be able to successfully retrieve relevant information. As a result, it is challenging to offer a unified method for preprocessing, indexing, and retrieving biomedical information.

Second, the mixture of natural English terms and biomedical-specific terms can pose problems due to high term ambiguity. A single word can have different meanings [3]. This ambiguity may in turn result in challenges for traditional IR methods, such as thesaurus-based extension of queries, as well as identifying relevant documents. For this reason, there is a strong need for word sense disambiguation, which is not easy, and is in itself an area of active research [3]. To illustrate, the term "*SOS*" normally refers to "*urgent appeal for help*", but it could also mean the gene symbol "*SOS*", short for "*Son of sevenless*"[1]. Another term illustrating this challenge is "*NOT*". It can refer to a protein[2]. But in traditional IR, it would be categorized as a stop word and, thus might be ignored in the indexing process.

Third, one of the problems with biomedical information is the lack of widely recognized terminology standards. New names and terms are created every time a biologist discovers a new gene or other important biological entities, and there often exist several inconsistent typographical/lexical variants [2]. Authors also normally have their own writing style [4], which can further worsen the situation.

Fourth, partly as a result of the existence of several term variants, important words and symbols suitable for indexing have often a low occurrence frequency, and many of terms appear only once in the entire document collection. While this may, in some cases, be useful since the discrimination effectiveness is often inversely proportional with the word's document frequency, it may also mean a high data sparseness, which would, in turn, have negative effects on the retrieval performance [5].

In summary, current biomedical IR systems have to deal with heterogeneous and inconsistent information. For static document collections, the above characteristic would be less problematic. However, document collections can be very dynamic. As shown in Figure 1[3], every day approximately 1000 new citation

---
[1] See for example http://www.ncbi.nlm.nih.gov/sites/entrez?db=gene&cmd=retrieve &dopt=default&rn=1&list_uids=6654
[2] See for example http://www.uniprot.org/uniprot/P06102
[3] This graph was generated based on the data at http://www.nlm.nih.gov/bsd/licensee/baselinestats.html

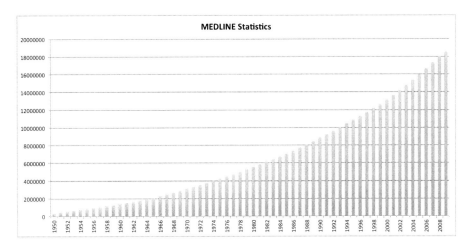

Fig. 1. Number of MEDLINE citations from 1950 to 2009

records are added to MEDLINE[4]. In addition, there is a challenge to measure document relevance to queries. While the notion of relevance based on, for example, the vector space model [6, 5, 7] is well-known and works very well for traditional IR, it needs improvements in biomedical IR, mainly because of the characteristics described above. As a result, the notion of relevance must be examined carefully, knowing that for a specific query, what is relevant for one biologist may not necessarily be relevant for another.

1.2 Objectives and Contributions

Our goal is to provide a generic biomedical IR system that can be *adapted* to meet different information needs. A basic requirement for such a system is that it must address the fact that users often have their own view of relevance. Moreover, the heterogeneity of biomedical information has made it important to have a customized system. This means a system that

(1) allows a user to specify relevant documents from the returned search results through user relevant feedback (URF), and
(2) allows us to specify the similarity/language model and the importance of specific parts in a document.

From this point of view, the main objective of this work is to investigate information retrieval methods that provide the best ranking of relevant results. More specifically, the main contribution of this work is improving existing typically keyword and parametric-based similarity models, such as the BM25 model,

---
[4] MEDLINE is a trade-mark of the U.S. National Library of Medicine
(See http://www.nlm.nih.gov/databases/databases_medline.html)

by enabling boolean queries and wildcard queries, and by applying a query-document-correlation factor to affect the ranking with respect to the amount of overlap between a query and a document (see also Section 3). We provide an extensive evaluation of our method based on known evaluation principles and a proof-of-concept prototype implementing the method. Although we recognize its usefulness, we show that we can achieve the improvement without using a controlled vocabulary such as the Unified Medical Language System (UMLS)[5] and the Medical Subject Headings (MeSH) [8].

1.3 Paper Outline

The remainder of this paper is organized as follows. In Section 2 we discuss what text operations can be used in preparation of index terms. In Section 3, we elaborate on our approach to find optimal scoring schemes and discuss the effects of applying boolean operations with BM25. In Section 4 we discuss how users can be involved in ranking of search results, through user relevant feedback and a fast, interactive user interface. Then, in Section 5 we present and discuss the results from testing the method and the prototype. Section 6 discusses related work. Finally, in Section 7 we summarize the paper and present future work.

2 Choosing Optimal Text Operations

Applying text operations in information retrieval is a process of choosing candidate words or tokens to index [5, 7]. Tokenization and stemming are examples of such operations. While tokenization is straightforward for general English text, this is not necessarily the case with biomedical information due to the characteristics described in Section 1.1. Existing empirical studies on tokenizing biomedical texts further emphasize this [9, 10].

Both studies suggest applying *conditional stemming* as part of the normalization process after tokenization. Stemming means that we only index the root or the *stem* of a specific term, while at the same time being able to search all grammatical variants of that term – e.g., `activate` versus variants such `activates`, `activated`, and `activating`. Based on the results from [9, 10], we believe that stemming would work if it is customized to biomedical texts. However, because the benefit versus cost of stemming is often debatable [5, 7], we chose to compare the effects of the use against omitting stemming. Our conclusion is that stemming did not give significant benefits in terms of retrieval performance. Further, for some types of queries it could change the semantic meaning of a query. For example, a query to find all proteins that are *activated* by a specific protein could also include a query finding all proteins that *activate* the actual protein. This would clearly be a different biological process. With conditional stemming we could take such semantic differences into account. However, this is beyond the scope of this work.

---

[5] See http://www.nlm.nih.gov/research/umls/index.html

So instead of stemming, we considered the use of *wildcard queries*. With a wildcard query, we can choose a specific query term as a prefix. Then, we extend the original query with all terms in the index that have the same prefix. For example, a query with `ferroportin*` will find articles with `ferroportin1` or `ferroportin-1`. This approach gives the users the freedom to choose when to extend the queries with the term variants. Our experiments also show its usefulness in terms of better retrieval performance.

We also investigated the use of stop-word removal as part of the text operations. This process removes words or terms that do not contribute to discriminating documents. Our experiments show that by carefully choosing candidate words in this process, we can achieve good results at the same time reduce the size of the index.

3 Finding Optimal Scoring Scheme towards Optimal Document Rankings

To further address some of the challenges in Section 1, we need to develop a good ranking method. Existing scoring schemes are good for general purposes, but they need improvements to meet our needs in biomedical information. When developing our method and prototype, we investigated several different schemes, and used them as a starting point towards a more improved scheme. In addition to designating suitable ranking models, we also try to find out how we can influence the ranking by boosting important parts in the document and constraining documents based on their query matching degree.

3.1 Choosing Ranking Models

Several ranking or similarity models have been proposed over the years. One of the earliest, proposed by Salton and Buckley [6], was the Vector Space Model (VSM). It is a similarity model based on cosine similarity between a specific query and the documents in the collection. Its main idea is that for all documents in the collection that contain the terms in the query, a ranking is produced by measuring the cosine of the angle between the query and the document vectors. For general text, it has been shown that this model gives good retrieval performance [5, 11]. Because VSM is one of the most used models, we applied it as a baseline in our experiments. In addition, we implemented an extension based on Lucene[6] [12].

Another similarity model that has been increasingly popular in information retrieval is the Okapi BM25 [13]. In contrast to the VSM, BM25 is a probability-based approach. As with VSM, it has been shown to be effective on general text retrieval [7]. This is also partly why we decided to implement it in this work. Moreover, we want to know how it performs when used in a highly scientific domain such as the biomedical domain. We first implemented the baseline model

---
[6] See http://lucene.apache.org/java/docs

as an extension to Lucene. Then, as shown below, we extended it to take into account that parts of the documents have different boosts, which will, as a result, also affect the way the search results are ranked.

Furthermore, in order to compare their performance in the biomedical domain, we implemented Lemur TF-IDF [14] and Divergence From Randomness BM25 (DFR-BM25) [15]. These are models developed as an alternative to the Okapi BM25 model [15, 14, 13]. Lemur TF-IDF is a variant of TF-IDF using a TF component that is similar to the one used in BM25. The Divergence From Randomness (DFR) BM25, on the other hand, is a probabilistic model in the same category as BM25. However, while BM25 uses some parameters to experimentally tune the model, DFR BM25 is non-parametric. As can be derived from the name, it assumes that the distribution of words in documents is a random process and that documents can be ranked by the probability that the actual term distribution found in a document would appear by chance.

Table 1. Summary of considered ranking models

Models	Description
VSM	A vector space model based on TF-IDF weighting scheme.
Okapi BM25	A parametric probabilistic model using TF, IDF, and document length.
Lemur TF	A variant of TF-IDF that uses the same TF parameters as in BM25.
DFR BM25	A non-parametric model using term probability distributions as a base for the result ranking.
Extended VSM	Extending the standard VSM model with query overlap factors, boolean operators, and wildcard queries.
Extended BM25	Extending the standard BM25 model with query overlap factors, boolean operators, and wildcard queries.

In addition to the above models, we have considered using language models (LM) for information retrieval [16, 11]. The basic idea of the language model is similar to the probabilistic model. However, instead of modelling the probability for a document to be relevant as with a probabilistic model, in a language model we build a probabilistic model for each document in our collection and rank our documents with respect to the probability of the model to generate the query. For this reason, this model would fit well with our goals on ranking of documents based on the user's interactions. Studies have also shown that the LM competes well with the other models with respect to retrieval performance [16, 11]. Nevertheless, due to time constraints its use and integration with BioTracer have been left for future studies.

3.2 Boosting Specific Document Parts

Boosting specific parts in a document is one way to specify how important one part in a document is compared to other parts. By defining this importance, we implicitly specify how the documents in the search results should be ranked. A document part can be (1) a structural part such as the title or the abstract, or (2) specific sentences or terms. Previous studies have shown that boosting structural parts of documents can improve the result quality [17, 18].

Structure-Based Boosting: By default, we perform search on both the *title* and the *abstract* fields of a document. These fields can be given boosts during index time or search time. A title can describe the content of a document well. Thus terms in a title can readily be assumed important, and given a higher weight than those in other places in the document. For instance, our experiments show that weighting the *title* field twice as much as *abstract* yields best ranking. Another advantage of this approach is that by having different weights for these fields, we also account for the fact that many MEDLINE documents may only contain a title and no abstract. With the same weight, documents without an abstract would automatically be ranked lower.

Sentence and Term-Based Boosting: We also investigated the effect of boosting sentences and terms during the index time. This has several advantages. First, it allows us to systematically specify which sentences or terms are to be weighted higher or lower than others. Second, specifying higher boosts or weights based on the importance of specific terms and sentences can help optimizing the ranking of retrieved documents.

The idea is to use techniques from the text mining and natural language processing (NLP) field, including named-entity recognition (NER) [19, 20] to identify special sentences and words. Such sentences could be those containing biological-specific names such as drug, gene and/or protein names, or biological processes such as protein-to-protein interaction and DNA evolution. The idea of identifying specific words or entity in texts is still an active research area [19, 20, 21] because of the diversity in biomedical text as discussed in Section 1.1. To our best knowledge, there has not been much work that has addressed boosting sentences and words as part of a ranking strategy. By successfully identifying biomedical specific words and sentences we can weight these higher than other general words and sentences. These weights will, in turn, contribute to ranking specific documents higher than other documents.

Our preliminary experiments in this area have shown promising results [22]. It is still in an early stage and for this reason, more work and experiments are needed to make it an integrated part of the current work.

Query-Time Term Boosting: In addition to index time boosting, specific terms can be given higher boosts than other terms during search time. The main advantage of this method is that terms that a user considers to be more important can be given an additional weight, in addition to their statistical information, such as term frequency (tf) and document frequency (df). As an example, assume we want to find all articles describing the function of the protein `ferroportin1`. We could compose a query as `ferroportin1^2.0 protein` function. This query tells the search engine that we would like to weight `ferroportin1` as twice as much as the words `protein` and `function`.

3.3 Constraining Documents Based on Their Query Matching Degree

Computing term weights: In general, a document can be given a score based on its similarity to the query:

$$Score(q, d_j) = sim(q, d_j) = \sum_{i \in q} w_{ij} \cdot w_{iq}. \tag{1}$$

Here $\vec{d_j} = (w_{1j}, w_{2j}, \ldots, w_{sj})$, $w_{ij} \geq 0$ is the vector of term weights for a document d_j, and $\vec{q} = (w_{1q}, w_{2q}, \ldots, w_{sq})$, $w_{iq} \geq 0$ is a query q, where s is the number of terms in the index.

Now, assume d_j consists of n fields f_{kj}, $k = 1, 2, \ldots, n$. Taking the boosting values into account, we have

$$\vec{d_j^*} = \sum_{k=1}^{n} \beta_k \cdot \vec{f_{kj}}, \tag{2}$$

where β_k denotes the boosting value for field f_{kj}. A special case can be defined as

$$\vec{d_j^*} = 2.0 \cdot \vec{f_{1j}} + 1.0 \cdot \vec{f_{2j}},$$

where f_{1j} and f_{2j} are the title and abstract fields, respectively.

Using the Okapi BM25 model [13], we assign each term in d_j and q the following weights w_{ij} and w_{iq}, respectively:

$$w_{ij} = \log\left(\frac{N - df_i + 0.5}{df_i + 0.5}\right) \cdot \frac{(k_1 + 1)tf_{ij}}{K + tf_{ij}} \text{ and } w_{iq} = \frac{(k_3 + 1)tf_{iq}}{k_3 + tf_{iq}}, \tag{3}$$

where $K = k_1((1 - b) + b(L_{d_j}/L_{avg}))$, N is the number of documents in the collection, df_i is the document frequency – i.e., the number documents in the collection that a term t_i occurs in, tf_{ij} is the frequency of a term t_i within d_j, L_{d_j} is the length of d_j – i.e., the number of terms in d_j, L_{avg} is the average document length, while k_1, k_3 and b are tuning constants with default values 1.2, 2.0. 0.75, respectively[7].

Now, let w_{ikj} be the weight of a term in f_{ij} of d_j^*. Then, applying weight on Eq. 3, we have [18]:

$$w_{ikj} = \log\left(\frac{N - df_t + 0.5}{df_t + 0.5}\right) \cdot \frac{(k_1 + 1)\beta_k tf_{ij}}{K + \beta_k tf_{ij}}, \tag{4}$$

where $K = k_1((1 - b) + b(L_{d_j}/L_{avg}))$.

---
[7] The values were set in accordance with the recommendation by Robertson and Jones [13].

Query-Document Correlation Factor: In the attempt to improve the ranking of relevant documents, we have added a new factor based on the degree of query matching. The idea is to add a scoring factor based on how well a document matches a query. We call this constraining documents based on their query matching degree, or simply *query-document correlation factor*. Such a constraining process is used in the literature. For instance, Lucene allows a form of document constraining too [12]. In our approach, however, we apply a slightly modified factor. In addition, we combine it with the similarity model BM25 [13]. To our best knowledge, using the idea of document query correlation factor with BM25 is new. Note that this factor does not directly help to increase the overall retrieval precision or recall, but it is meant to influence the ranking of already retrieved results. Its main goal is to maximize the number of relevant documents in the top-k hits of the search results.

Inspired by the scoring scheme in Lucene [12], we extended the original BM25 scoring equation by adding a document-query correlation factor, called $\Gamma(q, d_j)$, for a query q and a document d_j. This factor is calculated based on how many terms in q are found in d_j, and is proportional to the overlap between q and d_j. In our approach, $\Gamma(q, d_j)$ is computed as follows:

$$\Gamma(q, d_j) = \left(\frac{n_{over}(q, d_j)}{max_l \; n_{over}(q, d_l)} \right)^\theta = \left(\frac{\sum_{i=1}^{n} n_{over}(q, f_{ij})}{max_l \; n_{over}(q, d_l)} \right)^\theta \quad (5)$$

Here θ is a natural number, $n_{over}(q, d_j)$ is the number of terms in q that overlaps with document d_j. For all documents in the result set, $max_l \; n_{over}(q, d_l)$ is the maximum n_{over} for the retrieved documents.

Formally, $n_{over}(q, d_j)$ is defined as follows. Assume that t_i is a term index, $i = 0, 1, \ldots, s$, and s is the total number of terms in the index, i.e., $S = \{t_0, t_1, \ldots, t_s\}$ is the set of all distinct terms in the index. The overlap between a document d_j and q is defined as $\forall t \in S \mid t \in d_j \bigcap q$, $n_{over}(q, d_j) = |d_j \bigcap q|$.

We studied the effect of the factor Γ on the retrieval results and the overall evaluation result. We observed that with a small value of θ, the factor is too dominating. Thus it has unwanted effects on the ranking – i.e., relevant documents come too late in the result lists. Similarly, with a big value, θ will be too restrictive and the factor will have bad influence on the ranking. Therefore, the value of θ has to be chosen carefully. After several experiments, we found that $\theta = 4$ gave optimal results.

By combining Eq. 2, Eq. 4, and the extension of the similarity score from Eq. 1, we have:

$$Score(q, d_j) = \Gamma(q, d_j) \cdot sim(q, d_j^*) = \Gamma(q, d_j) \cdot \sum_{\forall i \mid t_i \in q} w_{ikj} \cdot w_{iq}. \quad (6)$$

3.4 Applying Boolean Queries with BM25

As for the Vector Space Model [6], BM25 was originally designed for keyword-based queries [13]. However, to be able to restrict the search results and filter out unwanted hits, we wanted to investigate the effect of using boolean operations in combination with BM25. Therefore, we implemented our prototype to allow users to search with boolean operations including AND, OR and NOT in combination with BM25.

The boolean query model was used in many search engines [5,7]. However, the original model has been seen to be too limited for several reasons. First, it does not rank retrieved results based on relevance to the query, and it only retrieves results that exactly match the query. To address this problem, we investigated the effects of combining boolean queries with BM25. The benefit is that, we can filter the search results based on the boolean operations in the query, and at the same time rank them based on the BM25 model. Second, a general criticism against boolean query usage is that users might be unwilling or even lacking the ability to compose this type of queries. While we recognize that this concern is valid, we still believe that a system has to provide the possibility of processing boolean queries. In fact, a PUBMED query log [23] with approximately 3 million queries shows that 36.5% of the queries were boolean queries. Although many of these queries might be computer generated (e.g. by automatic query extension), it is a strong enough motivation for investigating the effects of boolean queries. Moreover, although boolean queries do pose some challenges when it comes to finding all relevant documents from the collections, our experiments (see also Section 5.2) show that we can improve the average top-100 precision by 17% compared to the baseline BM25. This result means that boolean queries combined with BM25 can have good effect on the search precision.

4 Other Features

4.1 User Relevant Feedback (URF)

As suggested in our previous discussion, there are many ways to involve a user in influencing the ranking of documents. Involving the user through relevant feedback has been studied within traditional and multimedia information retrieval for decades. However, within the biomedical domain it seems still missing. A challenge here is to choose the right strategy. In our work, we have chosen to investigate the use of URF based on the scoring scheme suggested by Robertson and Sparck Jones [13]. In addition, we apply the extended BM25 as described earlier. That is, we also extended the original model with the document-query similarity factor $\Gamma(q,d)$. The scoring scheme equation is similar to the one in Eq. 6, but now the term weight includes the relevance information added by the users. Based on the ideas of Robertson and Sparck Jones [13] and Eq. 4, this weight can be expressed by

$$w_{ikj} = \log\left(\frac{(r+0.5)(N-n-R+r+0.5))}{((n-r+0.5)(R-r+0.5))}\right) \cdot \frac{(k_1+1)\beta_k \text{tf}_{ij}}{K + \beta_k \text{tf}_{ij}}, \quad (7)$$

where $K = k_1((1-b) + b(L_d/L_{avg}))$, r is the number of retrieved relevant documents, and R is the number of relevant documents in the collection for a specific query.

Here, the value of R can be derived from users' click-through information in the query log. Click-through information may work as relevance judgement information in that we can assume that all documents that a user has clicked are relevant to a specific query [7]. This means that we can estimate the number of relevant documents based on the number of documents that are known relevant to the user. Although R may in this case be much smaller than the real total number of relevant documents in the collection, this estimation can still be reasonable [13], presuming that these R relevant documents are within a subset of the set of all relevant documents.

To be able to evaluate this approach, we simulated the user relevant feedback process using the relevance information from the test collection.

4.2 Interactive Autocompletion

The primary goal of a search system is to provide best possible retrieval performance in terms of good search results such as high recall and precision values. However, we argue that providing a tool that is able to provide results within a reasonable time frame is crucial. Users often require an interactive system when performing their searches. For this reason, keeping the search time as low as possible should be an important goal for every retrieval system.

As a result, we studied how to make our system interactive. Our system has the following features. First, BioTracer provides a responsive and interactive autocompletion, which is activated by each keystroke. This feature can help users compose their queries. Furthermore, as opposed to autocompletion in existing systems, autocompletion in BioTracer not only provides possible prefix-based suggestions, but also is fault-tolerant. This means that it can find answers even if the prefix in a query has spelling mistakes. Figure 2 illustrates how this feature is implemented in BioTracer.

These features are useful since quite often a user only vaguely knows what she/he is looking for. Thus by allowing interactive browsing, we can allow a user to decide which suggested keywords should be the correct ones. To enable this feature, we implemented the system as follows (see also Figure 3).

Index term extraction. We use the index term dictionary to build the autocompletion index. To do this, we have to first extract all unique terms and their corresponding weights from the document collection index.

Building the index. To enable keystroke autocompletion, we use an n-gram index strategy. This strategy allows us to do matching against term prefixes. The maximal gram length is 20. As an example, let `protein` be a term that we want to index and assume that we use 2-grams. Then, our gram index would be: `_p`, `pr`, `ro`, `ot`, `te`, `ei`, `in`, in addition to `protein`.

Prefix and fuzzy queries. The use of n-gram indexes enables prefix queries and allows us to retrieve terms that share their prefixes with input sequences of characters. In addition to allowing higher fault tolerance, we also allow fuzzy queries. This feature allows us to retrieve terms that are similar to the typed one. The closeness is measured using the Levingston distance (also called edit distance).

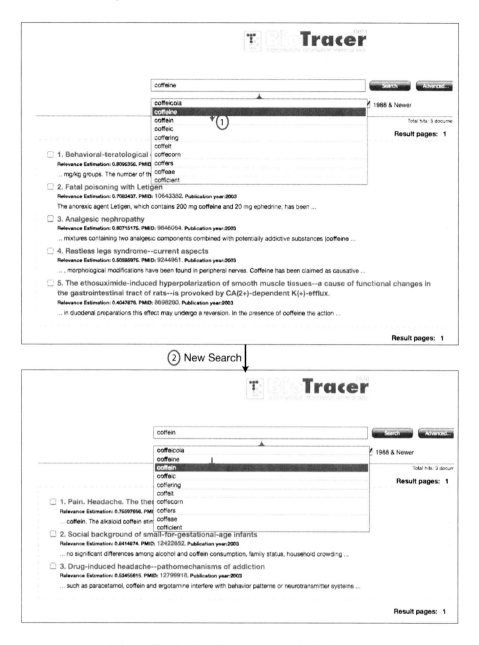

Fig. 2. Illustration of the autocompletion feature

Ranking. Once retrieved, the terms are ranked based on several factors. We mainly rank them based on their term weights stored in the main index. As a user types, words with most similar prefixes are readily retrieved first. Then, for terms that do not share prefixes, we use edit distance to order

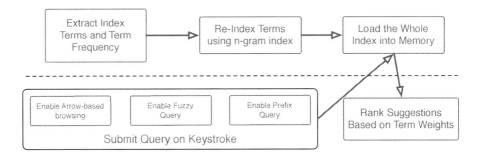

Fig. 3. Summary of the approach to achieve interactive autocompletion

terms that are related as specified by the fuzzy query. Term weights are the TF-IDF weight when we use the vector space model (VSM), while we use the weight in Eq. 4 for our BM25-based ranking model.

5 Evaluation

5.1 Prototype Implementation

In order to test our ideas, we implemented a proof-of-the-concept prototype, called BioTracer. Instead of re-inventing the wheel, we decided to use existing open source libraries to implement the prototype. Because of the requirements for performance, extensibility, simplicity, as well as scalability, we chose Java Lucene.

Figure 4 shows the BioTracer architecture. We use Lucene as a basis for indexing documents and handling queries, extended with a graphical user interface (GUI) based on JSP (Java Sever Pages) and AJAX (Asynchronous JavaScript and XML). The index is constructed mainly based on the MEDLINE database, which is parsed and handled by the Document Handler. To facilitate the parsing and generation of Lucene-friendly documents, we use Lingpipe[8] MEDLINE parser as a base for implementing the Document Handler. This handler also interacts with both the Language Model Handler and IndexManager to index documents based on a chosen language or similarity model.

All search is logged in the Search Log repository. This is used to "learn from experience", allowing BioTracer to use data from previous searches and user choices/interactions (e.g., URF). The Log is implemented using MySQL.

Figure 5 shows the implemented Web-based GUI, which also illustrates the URF possibility as described above. Each returned relevant hit can be marked by the user. The system will then use this information to re-rank the results.

Note that due to US National Library of Medicine (NLM) license regulations, BioTracer is not allowed to provide abstracts directly from MEDLINE. However, we have implemented a workaround that still allows users to browse the abstracts

---

[8] See http://alias-i.com/lingpipe/

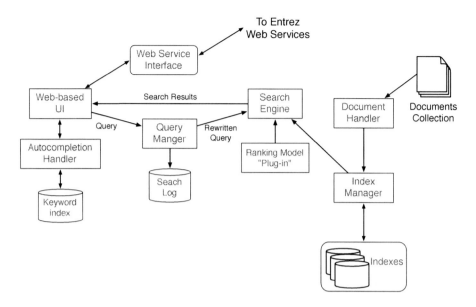

Fig. 4. The BioTracer architecture

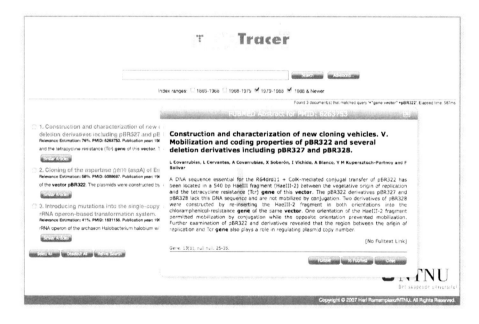

Fig. 5. Screen dump showing the BioTracer search environment

inside BioTracer. The solution is using the Entrez Web Services interfaces in combination with AJAX. This allows us to retrieve a specific abstract from PubMed based on the PubMed ID (if available) and process it. In this way, we are able to allow users to study the abstracts inside BioTracer, rather than going to another Web page. Moreover, if the user wants to examine more than one abstract, our AJAX implementation allows several PubMed abstract windows to be popped up at a time.

5.2 Experimental Results

Experimental Setup: To test the retrieval performance of BioTracer, we have performed the evaluation based on the TREC 2004 Genomic Track [24] test collection. The main reason for choosing this collection is its number of topics covered, which span from gene name searches to disease searches. It allows us to do a comprehensive evaluation of the BioTracer retrieval performance.

The TREC 2004 collection originally consists of 4,591,008 MEDLINE documents. Our evaluation was based on a subset of 42,255 documents that were judged against 50 topics. Each topic consists of *title*, *need*, and *context* fields. The queries were generated manually from the topics using all the fields. The runs were carried out as follows. First, we tested the generated queries against an off-the-shelf Lucene similarity model. Then, we used the same queries on the extended models that we discussed earlier.

To evaluate the retrieval performance, we used the same approach implemented for Text REtrieval Conference (TREC), that is, by focusing on TREC mean average precision (MAP). For each query, we may get the average precision by computing the average precision value for the set of top-k documents after each document is retrieved. Thus, MAP is the mean of the average precision values for all the queries. MAP has been widely used as the measure to evaluate ad hoc retrieval results [25,26]. It has also been used in the TREC Genomics Track evaluation [24]. Since MAP is sensitive to the rank of every relevant document, it reflects well the overall ranking accuracy. For this reason, the best result is the one with highest MAP values.

Further, since users are generally most interested in the results within the top 100 retrieved documents, we have stressed measuring the BioTracer's ability to find relevant data among the top 100 hits. Therefore, we measured specifically the precision at 10 retrieved documents (P@10) and precision at 100 (P@100).

To make our evaluation as independent as possible and to be able to compare it with previous results, we used the Buckley's **trec_eval** program[9]. Using trec_eval, the maximum number of allowed retrieved documents for each query was, by default, set to 1000. For each run, documents without any specific judgement were treated as non-relevant.

Results: Table 2 summarizes the results from our evaluation. The first column is the result from running the baseline method using the LuceneTFIDF-based

---
[9] See http://trec.nist.gov/trec_eval/

similarity model (VSM). The next column shows the results after we modified this model with the same extensions as those for BM25. Furthermore, the third column contains the result from running the baseline BM25 model. The result from our extensions in Sections 2, 3, and 3.4 are presented in the "Boolean+BM25" column, and the results from using user relevance feedback is shown in the "URF BM25" column. In addition to Table 2, Figure 6 shows the average precision values with respect to the number of retrieved documents and the used method. The different number of retrieved documents are represented by precision points. As an example, a precision point 20 means that we calculated the average precision at 20 retrieved documents.

Note that in addition to these results, we have tested the BioTracer prototype using Divergence From Randomness (DFR) BM25 [15], and Lemur TF-IDF models [14] with boolean and document boosts. However, the results from these runs were not significantly different from those of Boolean BM25. Therefore, they are not included in this paper.

Table 2. Results from running based on pooled TREC 2004 corpus

	Baseline Lucene TFIDF	Custom TFIDF	Baseline BM25	Boolean +BM25	URF BM25
MAP	0.346	0.4975	0.398	0.5122	**0.5129**
R-precision	0.3837	0.5443	0.4547	**0.558**	0.5527
P@10	0.534	0.652	0.594	0.666	**0.712**
P@100	0.3464	0.4552	0.3974	0.469	**0.4734**
Total recall	0.67	0.737	0.637	0.735	**0.738**

The results in Table 2 and Figure 6 show that there are only slight differences among the extended runs, although the URF BM25 had the overall best retrieval performance. These differences were most obvious for the top-15 hits. We can further see that the extended models performed much better than the baseline ones, which also show the effects of applying our extensions. For the TF-IDF model (i.e., the vector space model), the improvement of our custom TFIDF over the baseline model was 43.8%. The improvements for the BM25-based models were 28% and 29% respectively from the baseline model to the extended BM25 model and from the baseline model to URF BM25. Focusing on these results, adding the extensions is useful, in terms of retrieval performance. Further, comparing to the results from TREC 2004 Genomic Track [24] and assuming similar experiments, all of our extended models performed better than the best run from this track, where the MAP, P@10 and P@100 were to 0.4075, 0.604 and 0.4196, respectively [24].

Despite the above promising results, there are issues that we have to address. First, our evaluation was based on a somehow "closed" test collection. Although we have argued its comprehensiveness, we recognize that such a collection would hardly cover absolutely all aspects of users' needs in real search environments.

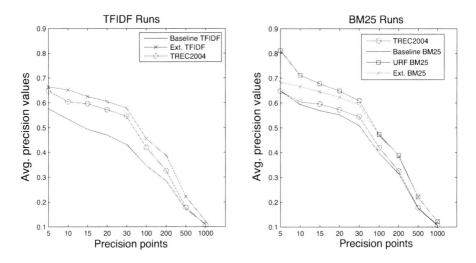

Fig. 6. Graphs for top-k precision values

This in combination with the fact that the BioTracer prototype can be seen as a highly interactive information search system, the generality/completeness and validity of the collection might be debatable. A way to make our evaluation more general is to involve real users, and study how they interact with BioTracer, as well as their evaluation of the retrieval performance. However, it is still not possible to guarantee a perfect evaluation [27], since we have to rely on several factors [28] such as the users' willingness to answer the interviews, the completeness of the questionnaire, and the broadness of the users.

Execution Speed: With respect to execution speed, although we argued the necessity of having an information retrieval system that can handle high-speed search (see Section 4.2), it has not been the main focus of this work. Rigorous performance tests, including comparison with other approaches, are needed to assess our system performance.

Still, our informal tests have shown promising results. We tried to run our tests using 4.5 Millions documents, a subset of MEDLINE entries on an Apple iMac machine with 2.93GHz Intel i7 CPU (quad core) and 8GB RAM. Using the aforementioned queries from the Pubmed one-day query log [23] (see Section 3.4), the search speeds were measured as follows. First, focusing on the speed as function of the document frequency, which varied from 1 document to 1.3 Millions documents and with one term in each query, the execution speed was constantly below 100 ms. Most specifically, the execution speed was 68 ms at maximum and 3.6 ms in average. Second, focusing on the speed as function of the number of query terms, our system managed to run our queries within 150 ms in average, when we varied our query lengths from 1 to 25 terms, whereas the maximum speed was 657 ms. Here, we got the lowest speeds when the number of queries were higher than 20. We believe the main reason for this is the way Lucene parses and process the queries.

In summary, the above results shows that our system is able to execute any query very fast independent of the number of returned results as long as the we keep the number terms in the query low (i.e., below 20 terms).

6 Related Work

There are many methods and systems suggested and developed in the biomedical information retrieval research domain. Most previous work has been presented in several TREC Genomic Track conferences, including the aforementioned TREC 2004 [24]. Other approaches applying tokenization heuristics were studied by Jiang and Zhai [10]. Their evaluation has shown that these methods can successfully improve the retrieval performance.

Like ours, many existing approaches have focused on expanding queries [24, 29]. However, we have attempted to go further by combining the use of different ranking models that can be adapted to different types of information needs, boosting different parts in the documents, applying a query-document correlation factor, using wildcard queries as query extension rather than stemming, using boolean queries with the BM25, and involving the users by user relevance feedback.

Concerning retrieval systems, there are several related systems. Because these systems mainly use proprietary ranking and indexing strategies, our discussion will focus on the system features and functionality.

First, perhaps the most used retrieval tool is PubMed[10]. PubMed provides several useful features for searching biomedical publications, such as the use of MeSH terms to expand queries and the possibility to access related articles. However, to our best knowledge, it does not provide ranking based on document relevance, per se. Instead, PubMed seems to use a proprietary algorithm to order the search results based on publication date, author names, and journals. Nevertheless, a more useful Web-based GUI extension to PubMed is available, called HubMed [30]. Like BioTracer, HubMed implements web service facilities for PubMed.

Another promising system is iPubmed [31]. The iPubmed system focuses on providing high performance and interactive search interface to MEDLINE data, through a "search-as-you-type" feature. It means that a search is initiated as soon as a keystroke is hit. Another useful feature of this system is their implementation of fuzzy search, allowing correct retrieval of documents even though users have misspelled their queries. Somehow, our autocompletion feature is inspired by this system with respect to instant and fuzzy search. The main difference is how the search results are retrieved and the ranking strategy applied.

Third, ScienceDirect is a search system[11]. It makes use of the powerful Scirus search facility. Scirus uses a more sophisticated ranking algorithm, which seems to produce more relevant results than PubMed. To our best knowledge, the main difference with BioTracer is that ScienceDirect/Scirus does not allow any

---

[10] See http://www.pubmed.org
[11] See http://www.sciencedirect.com

kind of user relevance feedback to refine the search. It is also worth noting that SienceDirect is a commercial system that covers a much larger scientific area than BioTracer. Therefore, the search results from this systems often include other than biomedical documents.

A fourth system worth discussing is Textpresso [32], which is also a search engine allowing specialized search of biomedical information. In contrast to Bio-Tracer, however, Textpresso extensively uses ontologies in the retrieval process. For a biomedical text, each word or phrase is marked by a term in the ontology when they are indexed. Its ranking algorithm is based on the frequencies of queried index terms. This means that the document containing the largest number of query terms is ranked at the top. To our best knowledge, Textpresso does not offer any user relevance feedback feature.

Fifth, BioIE [33] is a rule-based information retrieval system that extracts informative sentences from the biomedical literature such as MEDLINE abstracts. It uses MEDLINE as the main underlying searchable information. In addition, it allows users to upload their own text to be searchable. Both statistical analysis (word distribution, filtered word distribution, N-gram distribution and MeSH term distribution) and sentence extraction can be performed. BioIE extracts informative sentences based on predefined templates for a user-specified category. The ranking algorithms of BioIE are based on the numbers of occurrences of terms in the query, much like Textpresso. BioIE does not support user relevance feedback.

7 Conclusion and Future Work

In this paper we presented our ongoing work towards the development of an information retrieval system, called BioTracer. It is a search prototype that allows customized ranking of search results through boosting specific parts of documents, customizing scoring schemes, and allowing users to affect the ranking through user relevance feedback (URF). In summary, we have investigated the effect of using and/or extending existing models like TF/IDF-based models and BM25 with support for boolean queries, boosting documents and document parts, as well as enabling URF on retrieval performance and result ranking. Focusing on URF and the ability to customize the retrieval process, users are more flexible to specify their views of relevance. Thus we can address the challenges that we face in searching biomedical information. In this respect, the major contribution of this work is the development of the BioTracer search system providing adaptable searching and ranking of biomedical information. Although we built the system on existing techniques, to our best knowledge, the way we integrated them and extended them in one system is unique. Moreover, we have done experiments with the TREC collection to investigate the system's retrieval performance. We believe that these experiments on biomedical information, leading to our comparison of the different ranking models and their retrieval characteristics, are in itself an interesting result. Nevertheless, focusing on BioTracer as a whole, the main conclusion from these experiments is that BioTracer is a tool that is able to retrieve relevant information and that our extensions have helped improving the retrieval performance.

There are still challenges left for further studies. First, the system has only been tested against a TREC corpus. As discussed earlier, the strength and the validity of such a test can be improved. Therefore, we recognize the necessity of doing more empirical experiments with real users in a realistic environment. This additional work will reveal areas where our work can be improved. Further, we will further investigate the effect of including natural language processing (NLP) methods in the document handling process. This includes applying named-entity recognition (NER) to identify important keywords and sentences, thus further improving the way parts of a document are boosted. We are also investigating the use of controlled vocabularies such as MeSH [8] in both for query term boosting and extending queries. Finally, we will include a text classification facility. Our preliminary experiments based the Support Vector Machine (SVM) have, by far, shown good potential in terms of increased search precision. However, more work is needed to make such approaches efficient enough so that it does not degrade the overall system performance.

Acknowledgements. The author would like to thank Jon Olav Hauglid, Roger Midtstraum, Reidar Conradi, and Min-Yen Kan for their suggestions to improve this paper.

References

1. Ramampiaro, H.: BioMedical information retrieval: The bioTracer approach. In: Khuri, S., Lhotská, L., Pisanti, N. (eds.) ITBAM 2010. LNCS, vol. 6266, pp. 143–157. Springer, Heidelberg (2010)
2. Krauthammer, M., Nenadic, G.: Term identification in the biomedical literature. Journal of Biomedical Informatics 37(6), 512–526 (2004)
3. Chen, L., Liu, H., Friedman, C.: Gene name ambiguity of eukaryotic nomenclatures. Bioinformatics 21(2), 248–256 (2005)
4. Netzel, R., Perez-Iratxeta, C., Bork, P., Andrade, M.A.: The way we write. EMBO Reports 4(5), 446–451 (2003)
5. Baeza-Yates, R.A., Ribeiro-Neto, B.: Modern Information Retrieval. Addison-Wesley Longman Publishing Co., Inc., Boston (1999)
6. Salton, G., Buckley, C.: Term-weighting approaches in automatic text retrieval. Information Processing and Management 24(5), 513–523 (1988)
7. Croft, B., Metzler, D., Strohman, T.: Search Engines: Information Retrieval in Practice, 1st edn. Addison-Wesley, Reading (2009)
8. Lowe, H.J., Barnett, G.O.: Understanding and using the medical subject headings (MeSH) vocabulary to perform literature searches. JAMA 271(14), 1103–1108 (1994)
9. Trieschnigg, D., Kraaij, W., de Jong, F.: The influence of basic tokenization on biomedical document retrieval. In: Proceedings of the 30th Annual International ACM SIGIR Conference on Research and Development in Information Retrieval, SIGIR 2007, p. 803 (2007)
10. Jiang, J., Zhai, C.: An empirical study of tokenization strategies for biomedical information retrieval. Information Retrieval 10(4-5), 341–363 (2007)

11. Baeza-Yates, R.A., Ribeiro-Neto, B.: Modern Information Retrieval, 2nd edn. Addison-Wesley Longman Publishing Co., Inc., Boston (2011)
12. Hatcher, E., Gospodnetic, O.: Lucene in Action. Manning Publications Co., 209 Bruce Park Ave., Greenwich, CT 06830 (2005)
13. Robertson, S.E., Jones, K.S.: Simple proven approaches to text retrieval. Technical Report 356, University of Cambridge (1994)
14. Zhai, C.: Notes on the lemur TFIDF model. note with lemur 1.9 documentation. Technical report, School of CS, CMU (2001)
15. Amati, G., Rijsbergen, C.J.V.: Probabilistic models of information retrieval based on measuring the divergence from randomness. ACM Transactions on Information Systems 20(4), 357–389 (2002)
16. Ponte, J.M., Croft, W.B.: A language modeling approach to information retrieval. In: Proceedings of the 21st Annual International ACM SIGIR Conference on Research and Development in Information Retrieval, SIGIR 1998, pp. 275–281. ACM, New York (1998)
17. Wilkinson, R.: Effective retrieval of structured documents. In: Proceedings of the 17th International ACM SIGIR Conference on Research and Development in Information Retrieval (SIGIR 1994), pp. 311–317. Springer-Verlag New York, Inc., New York (1994)
18. Robertson, S., Zaragoza, H., Taylor, M.: Simple BM25 extension to multiple weighted fields. In: CIKM 2004: Proceedings of the Thirteenth ACM International Conference on Information and Knowledge Management, pp. 42–49. ACM, Washington, D.C., USA (2004)
19. Cohen, A.M., Hersh, W.R.: A survey of current work in biomedical text mining. Briefings in Bioinformatics 6(1), 57–71 (2005)
20. Leser, U., Hakenberg, J.: What makes a gene name? Named entity recognition in the biomedical literature. Briefings in Bioinformatics 6(4), 357 (2005)
21. Kabiljo, R., Clegg, A., Shepherd, A.: A realistic assessment of methods for extracting gene/protein interactions from free text. BMC Bioinformatics 10(1), 233 (2009)
22. Johannsson, D.V.: Biomedical information retrieval based on document-level term boosting. Master's thesis, Norwegian University of Science and Technology (NTNU) (2009)
23. Herskovic, J., Tanaka, L., Hersh, W., Bernstam, E.: A day in the life of PubMed: Analysis of a typical days query log. Journal of the American Medical Informatics Association 14(2), 212–220 (2007)
24. Hersh, W.R., Bhupatiraju, R.T., Ross, L., Roberts, P., Cohen, A.M., Kraemer, D.F.: Enhancing access to the bibliome: the trec 2004 genomics track. Journal of Biomedical Discovery and Collaboration 1(3), 10 (2006)
25. Yilmaz, E., Aslam, J.A.: Estimating average precision when judgments are incomplete. Knowledge and Information Systems 16(2), 173–211 (2008)
26. Voorhees, E.M.: On test collections for adaptive information retrieval. Inf. Process. Manage. 44(6), 1879–1885 (2008)
27. Käki, M., Aula, A.: Controlling the complexity in comparing search user interfaces via user studies. Information Processing and Management 44(1), 82–91 (2008); Evaluation of Interactive Information Retrieval Systems
28. Kelly, D., Harper, D.J., Landau, B.: Questionnaire mode effects in interactive information retrieval experiments. Information Processing and Management 44(1), 122–141 (2008); Evaluation of Interactive Information Retrieval Systems
29. Abdou, S., Savoy, J.: Searching in Medline: Query expansion and manual indexing evaluation. Information Processing & Management 44(2), 781–789 (2008)

30. Eaton, A.D.: Hubmed: a web-based biomedical literature search interface. Nucleic Acids Research 34(Web Server issue), W745–W747 (2006)
31. Wang, J., Cetindil, I., Ji, S., Li, C., Xie, X., Li, G., Feng, J.: Interactive and fuzzy search: a dynamic way to explore medline. Bioinformatics 26(18), 2321–2327 (2010)
32. Muller, H.M., Kenny, E.E., Sternberg, P.W.: Textpresso: an ontology-based information retrieval and extraction system for biological literature. PLoS Biol. 2(11), e309 (2004)
33. Divoli, A., Attwood, T.K.: BioIE: extracting informative sentences from the biomedical literature. Bioinformatics 21, 2138–2139 (2005)

Electronic Health Record Data-as-a-Services Composition Based on Query Rewriting

Idir Amine Amarouche[1], Djamal Benslimane[2], Mahmoud Barhamgi[2], Michael Mrissa[2], and Zaia Alimazighi[1]

[1] Université des Sciences et de la Technologie Houari Boumediene
BP 32 El Alia 16111 Bab Ezzouar, Alger, Algeria
`I.A.Amarouche@gmail.com, alimazighi@wissal.dz`
[2] Université de Lyon, CNRS
Université Lyon 1, LIRIS UMR5205
43, bd du 11 novembre 1918, Villeurbanne, F-69622, France
`{firstname.lastname}@liris.cnrs.fr`

Abstract. Due to the large development of medical information systems over the last few years, there is today a strong need for an infrastructure that uniformly integrates the distributed and heterogeneous collections of patient data to deliver value-added information to healthcare professionals at the points of care. The adoption of Electronic Health Records (EHRs) and Web services as a software infrastructure has become an extremely important prerequisite for patient data integration. In this paper we propose a semantic-enabled architecture for the automatic composition of EHR (Electronic Health Record) DaaSs (Data-as-a-Service). In our architecture, DaaSs are selected and composed automatically to resolve the user queries (i.e. queries posed by physicians, nurses, etc) using a query rewriting approach. Our proposed approach can also handle the semantic conflicts of data exchanged among component services in an EHR DaaS composition by deriving and applying automatically the necessary data conversions.

Keywords: Electronic Health Record (EHR), Data as a Service (DaaS), query rewriting, semantic annotation, composition, mediation.

1 Introduction

Our current health environment is characterized by a shared and distributed localization of patient information. Patients' data are spread across several autonomous, proprietary and heterogeneous information systems. The adoption of electronically formatted patient data with Electronic Health Records (EHR) has become the primary concern for a broad range of health information technology applications and practitioners. According to Healthcare Information and Management Systems Society [22] the Electronic Health Record (EHR) of an individual consists of a collection of lifetime health data in electronic format, generated during relevant interactions with the healthcare system.

In this context, one of the big challenges of the health actors is the communication, the sharing (via exchange and integration) of EHR data through several independent and heterogeneous health systems. Thus, the EHR data need to be available, discoverable, searchable and comparable by a connected group of care providers and health organizations.

For this reason, many efforts have been undertaken to identify the requirements and information architectures needed to support shared Electronic Health Records. These research projects have focused on supporting the care given to patients by promoting good designs for EHR systems and standards for the secure communication of part or all of a patient's EHR between authorized systems [4].

Also, many e-health systems already provide the possibility to export their data in standardized formats such as CEN TC251, openEHR, and HL7-CDA [3] which provide different ways to structure and markup the EHR data for exchange purpose. For this, the adoption of standardized Electronic Health Records has become an extremely important prerequisite for bringing interoperability and effective data integration to the healthcare industry [22].

However, these EHR-related standards have centered on the communication of parts of the EHR of an individual subject of care and deal with the patient data at the document level. Furthermore, many of the users' query requirements target the contents of the clinical documents [26] and little work has been done to date on defining a generic means of querying EHR systems, as distributed repositories, in a consistent way [4].

Today there is increasing interest in moving towards a Service-Oriented Architecture for EHR data sharing among independent health information systems [28] [19] [23] [22]. Web service technology can be used as a standardized way for accessing and sharing the EHR data over healthcare information systems. This type of services is known as Data-as-a-Service (DaaS). DaaSs allow for a query-like access to organizations' data sources and do not change the state of the world [34] [37]. In this paper we use the term *EHR DaaS* to denote the DaaS that provides EHR data (or parts of).

While individual DaaSs may provide interesting medical information alone, in most cases, users' queries require the composition of multiple DaaSs. Furthermore, as there are several EHR DaaSs provided by several health actors, the user (e.g. physicians) needs an assistance to discover, select and compose the required EHR DaaSs. Therefore a solution is needed to select and compose EHR DaaS automatically for the purpose of retrieving and integrating the EHR data, which is the problem we tackle in this paper. This problem is very complex since EHR DaaSs are developed by independent organizations that may use different standards to describe their data.

To address these challenges, we propose a declarative query-rewriting based approach for the automatic composition of EHR DaaSs. The key idea behind the approach is to describe DaaSs as views over medical ontologies to capture their semantics in a declarative way. Defined views are then used to annotate the EHR DaaSs' description files (e.g. WSDLs) and exploited to compose DaaSs automatically. They are also used in resolving (on the fly) the semantic conflicts of data exchanged inside DaaS compositions. Our composition framework is based on an RDF query rewriting algorithm [6] inspired by the mature research work done in the data integration area [17].

The rest of paper is organized as follows: Section 2 provides a motivation example, highlights the challenges addressed in this paper and describes our contribution. Section 3 provides some background knowledge about EHR data and EHR DaaS. Section 4 outlines our service-oriented approach for EHR DaaSs composition. We also present in this section our results in terms of models, which include an ontology model, a model for services (EHR DaaS and mediation services) and a conjunctive query model. In Section 5, we introduce the generic algorithmic solution for query processing, which includes a query rewriting approach for EHR DaaS composition and the automatic invocation of mediation services for the resolution of semantic conflicts. Section 6 shows the system implementation and Section 7 is devoted to related works. Section 8 summarizes the results obtained in this work and discusses some possible extensions.

2 Motivation, Challenges and Contributions

In this section, we provide an illustrating example where the information needs of health actors are satisfied with a service oriented approach. This approach raises up many problems, which motivate our proposal to apply semantic Web technologies to support EHR DaaSs composition.

2.1 Motivation Example

Let us consider an e-health system exporting the set of EHR DaaSs in Table 1 to query the patient data. The description of EHR DaaS can be seen in Table 1, where the symbols "$" and "?" denote inputs and outputs of EHR DaaSs, respectively. We assume that a physician wants to consult the laboratory test results for his patient, "Joe, 35 years old man". Laboratory test results are helpful tools for evaluating the health status of an individual. In each laboratory test order, we find several tests (Cholesterol rate, Bilirubine rate, etc...). The physician submits the following query, as shown in Figure 1: Q_1: "What are the pathologies indicated by the results of the laboratory tests of Joe"

For the sake of simplicity, we assume that the e-health system does not provides functionality (service location record) to find any EHR DaaSs providing Joe's health data. Doing so, the physician has to invoke the EHR DaaS that provides the recent laboratory tests ordered for Joe namely S_{11} or S_{12}. He will not invoke S_{13} or S_{14} which return the ordered test made by Gynecologist and Paediatrician specialist respectively. Also he will not invoke S_{15} because this service returns the laboratory tests ordered but aborted for a patient. After invoking S_{11} and S_{12} he will obtain the list of recent and successful laboratory tests ordered for Joe and the laboratories' names in charge for performing the tests. Then he will invoke S_{21} and S_{22} to retrieve the results of the laboratory tests ordered for Joe and made by *laboratory* 1 and *laboratory* 2 respectively.

Table 1. Example of Electronic Health Record Data-as-a-Services

Service	Functionality	Constraints and DaaS provider	The employed health standard
$S_{11}(\$x, ?y)$	Returns laboratory tests y ordered for a given patient x	DaaS provider is hospital1	
$S_{12}(\$x, ?y)$		DaaS provider is hospital2	
$S_{13}(\$x, ?y)$		Patient gender (woman), DaaS provider is maternity hospital	
$S_{14}(\$x, ?y)$		Patient age (< 14), DaaS provider is paediatric private hospital	
$S_{15}(\$x, ?y)$	Returns laboratory test y aborted for a given patient x	DaaS provider is hospital1	
$S_{21}(\$x, ?y, ?z)$	Return the name y and the value z of a given test belonging to lab test order x	z.unit (unit of measure) is mg/l, DaaS provider is laboratory 1	$y.code \in \{LOINC\}$
$S_{22}(\$x, ?y, ?z)$		z.unit (unit of measure) is mmol/l, DaaS provider is laboratory 2	$y.code \in \{SNOMED\}$
$S_{23}(\$x, ?y, ?z)$		DaaS provider is laboratory 3	
$S_3(\$x, ?y, ?z)$	Returns low reference value y and hight reference value z for given lab test x	z.unit and y.unit (unit of measure) is mg/l	$x.code \in \{SNOMED\}$
$S_4(\$x, ?z)$	Returns indicated disease z for abnormal value of lab-test x		$z.code \in \{ICD\}$ and $x.code \in \{SNOMED\}$

Furthermore, according to his own experience, the physician will not invoke S_{23} because of the inferior quality test results returned by the *laboratory* 3. After that, he will invoke S_3 which returns the reference or normalized values for each laboratory test parameter, in order to compare with the values returned by S_{21} and S_{22}. Then, if he found any suspicious values he invokes S_4 to retrieve the pathology indicated by each abnormal value. The list of pathologies returned by S_4 will indicate to him the pathologies of Joe may suffer from, and for which a treatment must be applied or another investigation is needed.

It is necessary to mention that during the comparison between the values of the test results returned by S_{21} or S_{22} and the references values returned by S_3, the physician must operate a conversion between value units (mmol/l and mg/l)[1] because each piece of data is interpreted differently. Also, the physician may need to convert exchanged data between selected services. For example, he has to change the laboratory test code returned by S_{21} (codified using the LOINC standard) to codes acceptable by S_3 (codified using the SNOMED standard)

In short, the physician needs to discover and select services, to invoke them in a certain order, to make sure that the parameters of the services are compatible, to consolidate the results returned by each EHR DaaSs and to manually perform an ordered set of operations like joins, selections and projections.

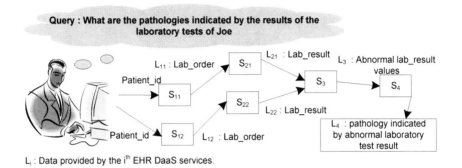

Fig. 1. Physician query scenario

2.2 Challenges

As shown previously, the manual process of composing EHR DaaSs for answering a query is painful and tedious; it may not possible for non expert users (eg. physicians, users...) to compose EHR DaaSs. Thus, automating the composition of EHR DaaS raises the following challenges:

1. *Understanding the Electronic Health Record Data-as-a-Service Semantics*: For the physician, confusion occurs in correctly understanding the functionalities provided by several EHR DaaSs. For instance, EHR DaaS like S_{11} and S_{15} have the same signature (input, output) but provide different functionalities, the former provides the laboratory test ordered for patient, whereas the latter provides the laboratory tests aborted for patient. Also, with several similar EHR DaaSs that offer the same functionality (like S_{11} or S_{12} or S_{13}) but that are subject to different constraints on data (patient age, patient gender,...), the physician must manually choose the ones that contribute to answering the query. Thus, the semantic annotation of EHR DaaS description files with the relationship between input and output on the one hand, and its content on another hand, will contribute mainly to automatic processing of EHR DaaSs discovery and composition and will spare the physician from selecting between several EHR DaaSs manually.

---

[1] In laboratory test results, mmol/l and mg/l are mass concentration unit measures.

2. *Electronic Health Record Data-as-a-Service discovery and composition*: The physician must select the services that are relevant to his query and compose them in the right order.
3. *Data level conflicts detection and resolution in Electronic Health Record Data-as-a-Services composition*: EHR DaaS parameters use concepts from different health ontologies such as SNOMED[2], ICD[3], LOINC[4], UMLS[5], ICPC[6] ...etc. For instance, S_{21} provides laboratory test coded in LOINC, but S_3 requires the laboratory test result specified using SNOMED terminology. Also, S_{22} returns test result measured in $mmol/l$ but S_3 measured its test result in mg/l. Consequently, the data will be passed during composition from S_{21} and S_{22} to S_3 will provoke an incompatibility problem. Then, how to leverage semantic conflict (health ontologies concept, unit,....etc) to enable the unrestricted composition of EHR DaaSs in a generic EHR environment becomes another challenge.
4. *Data trustworthiness*: Beside the semantic description of EHR DaaS, the physician need to have a mean to ensure the consistency and trustworthiness of data that are returned by different EHR DaaSs. Trustworthiness depends a lot on where data came from and which parties were involved in the establishment, change and forwarding of the data.

2.3 Contributions

Among the above mentioned challenges, we will focus only on the first three ones and we let the last one for future work. In a nutshell, we propose a new approach to integrate EHR data provided by several EHR DaaSs using a two-layer mediated ontology. The first layer, named *Generic Domain Ontology*, forms the core ontology (core concepts and relations) and it is the same for all EHR DaaSs providers. This layer is used to automate the composition of EHR DaaSs. The second layer, named *Specific Domain Ontology*, represents the (contextual) semantics (measuring units, scales, etc) of the data attached to EHR DaaS (called input/output parameters) and is used to detect and resolve the semantic conflicts of data exchanged among composed services.

The use of this two-level ontology allows deriving automatically the EHR DaaSs compositions that incorporate necessary mediation services (to carry out data conversion between interconnected services) during EHR DaaS composition. Our main contributions in the paper are summarized as follows:

– Firstly, we handle the first challenge by proposing an RDF-based modeling for EHR DaaSs. Specifically, we represent EHR DaaSs as *RDF views* over a mediated ontology. RDF views allow capturing the semantics of a DaaS in

---

[2] SNOMED: The Systematized Nomenclature of Medicine.
[3] ICD: International Codification of Disease.
[4] LOINC: Logical Observation Identifiers Names and Codes.
[5] UMLS: Unified Medical Language System.
[6] ICPC: The International Classification of Primary Care.

a declarative manner using concepts and relations whose semantics are formally defined in ontologies. We adopt SPARQL, the de facto query language for the Semantic Web, for posing queries over EHR DaaS services.
- Secondly, we propose a query rewriting approach to automatically select and compose the EHR DaaSs. In our approach composition queries, specified as SPARQL queries over a mediated ontology, are reformulated in terms of available EHR DaaSs based on the defined RDF views. Query reformulations are then translated into composition plans (i.e. orchestrations) that can be executed to answer the posed queries.
- Thirdly, since the data provided and required by individual EHR DaaSs may be bound to different (contextual) semantics (e.g. measuring units, scales, etc), we propose a mechanism that inserts automatically mediation services in compositions in order to resolve the semantic incompatibilities detected in the generated EHR DaaS compositions.

3 Background

In this section, we describe the features of EHR data published by EHR DaaSs. We also look at the relevant aspects of EHR data integration using Web service technology and the standards that EHR systems utilize nowadays. Furthermore, we look at the query rewriting approach for view based data integration, which is utilized in this work for EHR DaaS composition.

3.1 Electronic Health Record Standards

There have been various definitions of EHRs. According to the Healthcare Information and Management Systems Society[7] "EHR is a longitudinal electronic record of patient health information generated by one or more encounters in any care delivery setting". A comprehensive survey and analysis of the electronic healthcare record is available in [14].

EHR data is stored in many kinds of systems and proprietary formats, inducing different internal structures. This situation leads to severe interoperability problems in the healthcare informatics domain. For this purpose, several EHR standards have been developed in order to structure the EHR data for the purpose of exchange. The standardization effort focused around two considered areas.

- The structuralization of EHR documents for the exchange of clinical data by supporting meaningful information representation between clinical information system within or between health care organizations. These standards include Health Level 7 (HL7) Clinical Document Architecture (CDA) [13], openEHR [35] and Cross-Enterprise Document Sharing (XDS) integration profile IHE[8], .. etc.

---
[7] HIMMS, http://www.himss.org
[8] IHE, "Integrating the Healthcare Enterprise" http://www.ihe.net.

- The health ontology used to represent the EHR data. To name a few examples of health ontology that model parts of the medical domain : ICD, LOINC, SNOMED, UMLS,..etc.

In sum, these approaches focus on the way of accessing the data rather than standardizing the data itself. A common feature of all emerging EHR standards is that the clinical concepts are modeled and expressed independently from how the data is actually stored in underlying data source. This challenge necessitated to:

- Select an appropriate technological infrastructure for making EHR data available at the point of care when authorized users need. In this context, Web services technology has been largely applied in healthcare domain by encapsulating existing EHR data within the Web service model and providing access to clinical data in a standard way.
- Adopt a novel modeling approach namely two-level modeling[14][7]. Two-level modeling approach in EHR system development divides the EHR data models into two separate ones. A generic information model and domain knowledge model. The domain knowledge model contains a set of constraints model ((simple and complex type), internal consistency (type, interval values, scale, unit, range), Reference Data (XML Format, health ontology)) on instance of the generic model entities.

3.2 Electronic Health Record Data Integration and Web Services

As explained previously, the introduction of Web service technology is motivated by the need to encapsulate the patient data in case of a EHR data exchange with another organization to perform a specialized medical procedure or for continuation of care. During this exchange, the execution of many operations are performed on EHR data before their exchange that concern (see figure 2).

- Retrieving relevant authorized patient data from the health organization information system, for instance : "Problem list", "past illness", "medication use", "present illness", "Family history", "Past surgical", "allergies",..etc.
- Coding this data using numerous standards that support ontological control at instance and type level by interlinking such health ontology (ICD-10, SNOMED,...) with the data definitions in standardized EHR documents (HL7,..etc) [24]. Using the coded data for creating a EHR document complaint XML clinical model (HL7/CDA, ...etc) ;
- Sending this document as SOAP encoded message to an appropriate destination.

However, the problem of EHR data integration is central in these systems. These latters deal with patient data at the document level, but health care data usage often is data centric, meaning that data should be extracted from various documents and then integrated according to specific criteria. As depicted by [24], even with a service approach, many interoperability problems still arise during EHR data integration.

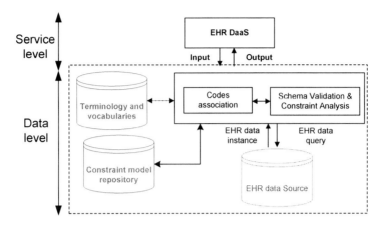

Fig. 2. Electronic Health Record data, publication through Electronic Health Record Data-as-a-Service

First, matching clinical data to codes in concurrent and semantically overlapping health ontology causes severe interoperability problems. Second, the semantic integration of heterogeneous systems in healthcare will have to deal with volatile medical concepts. For this reason, there is not, until now, a unique and comprehensive ontology of the medical domain [1] [30] [24].

In order to solve this problem, EHR data integration systems have to consider two levels; (1) generic information model and (2) domain knowledge (vocabularies, terminologies,..). These two levels must inter operate to integrate EHR data from disparate healthcare systems [33].

As every standard has its information reference model upon which domain knowledge is built, we will focus only on features of domain knowledge where EHR complaint XML documents at entry or section level are characterized by the frequent association with:

- Ontological concepts defined in some health ontologies (ICD, LOINC, SNOMED) for coding EHR;
- Semantic and structural constraint for maintaining internal consistency of EHR data;

Thus, the additional knowledge provided by the reference models upon which domain knowledge is built will not be addressed in this work.

3.3 Query Rewriting

The query rewriting problem has been extensively studied in the areas of query optimization and data integration. In the hereafter we report some definitions (based on the works [17,18]) to terms we use extensively throughout the paper.

Query Rewriting: Given a query Q and a set of view definitions $V = \{V1, ..., Vm\}$, the query Q' is an rewriting of Q using V if: Q' is contained in Q and Q' refers only to the views in V.

Query Containment and Equivalence: Query containment and equivalence enable comparison between different rewriting of a query. It will be used when we test the correctness of a rewriting of a query in terms of a set of views. Thus, a query $Q1$ is said to be contained in a query $Q2$, denoted by $Q1 \subseteq Q2$, if for all databases D, the set of tuple computed for $Q1$ is a subset of those computed for $Q2$, i.e., $Q1(D)$ and $Q2(D)$. The two queries are said to be equivalent if $Q1 \subseteq Q2$ and $Q2 \subseteq Q1$ or $Q1 \equiv Q2$.

Maximally Contained Rewriting: Equivalent rewritings may not always exist under the open world assumption. Finding the maximally-contained rewriting will be the only alternative for resolving a query. Thus, Let Q be a query, $V = \{V1, ..., Vm\}$ be a set of view definitions, and L be a query language. The query Q' is a maximally-contained rewriting of Q using V with respect to L if: Q' is a query in L that refers only to the views in V, Q' is contained in Q, and there is no rewriting $Q1 \in L$, such that $Q' \subseteq Q1 \subseteq Q$ and $Q1$ is not equivalent to Q.

4 Overview of the Approach

In this section we define our reference architecture for EHR DaaSs composition that is independent from specific data standards.

4.1 General Architecture

Our reference architecture defines four logical tiers, as shown in Figure 3.

- **Data Level:** The lowest level of the architecture contains information stored in different components. These components can be databases that store all the medical information concerning patients or documents that preserve all official documents generated during healthcare process. Also there are several clinical terminological and documentary resources that provide means to search and share clinical knowledge.

- **Service Level:** The service level publishes the different services provided by several systems to e-health actors. Services are either simple (one provider) or complex (multiple providers). This level provides two services categories.
  1. **Electronic Health Record Data-as-a-Services** provide information about patients. We can find two kinds of EHR DaaS in this category according to the nature of the data provided: EHR DaaSs that provide specific patient information (diseases, symptoms, medications or family history and so on) or EHR DaaSs that retrieve a clinical document complaint model (discharge summaries,...).

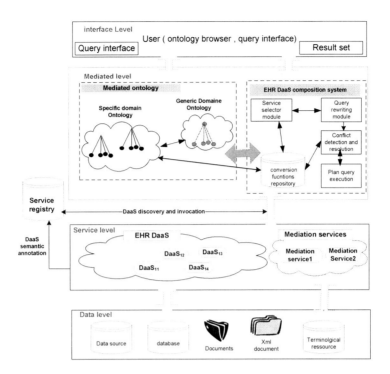

Fig. 3. Overview of the Electronic Health Record Data-as-a-Service composition architecture

2. **Mediation Services** is used mainly for mapping and converting the output parameter of a specific EHR DaaS to the input parameter of a subsequent EHR DaaS during service composition. Mediation services provide the definition of cross-mappings between terminologies (e.g; UMLS Terminology Services) and extract from a EHR DaaS output parameters what the inputs of subsequent EHR DaaS need.

These services advertise their WSDL definitions into a service registry. WSDL provides an XML-based grammar for describing a service interface. For automatic discovery, selection and composition of service, WSDL files are annotated with semantic entities from a mediated ontology. The service registry includes a set of services descriptions (WSDL files) semantically annotated with RDF views expressed in term of mediated ontology as in [6].

- **Mediated Level:** The mediated level is composed of two modules:
  1. *Mediated ontology*: The mediated ontology contains all the concepts and relations defined in EHR domain. It will be used to annotate and query services (EHR DaaSs, mediation services) in an environment of heterogeneous inter-working health information systems. We divide the ontology into two ontological levels which cuts the concept space into a

generic domain ontology and a set of extensions named domain specific ontologies.

   1-1. The Generic Domain Ontology (**GDO**) defines the generic concepts and relations covered the EHR domain and it is the same for all EHR DaaSs providers. For example, patients, disease, laboratory test are entities of the generic domain ontology. The generic ontology is different from a specific ontology in that it only contains basic shared concepts and their properties. For example, the generic ontology states that a `laboratory-test-result` has a unit and a code, without specifying any specific unit or health ontology code (this information will be specified using the SDO detailed later). Doing so, the GDO is used mainly for EHR DaaS discovering and composition.
   1-2. The Specific Domain Ontology (**SDO**) is created mainly to allow the semantic extension of generic ontology concepts for detecting and resolving semantic data conflicts. For example, the SDO states that a `Laboratory-test-code` or `Disease-code` are expressed in SNOMED or LOINC; or `laboratory-test-value` has `mg/l` as a unit of measure.

2. *Electronic Health Record Data-as-a-Service composition system:* Contains four sub-modules: a service selector module, a query rewriting module, a conflict detection and resolution module and a query plan execution module. The first module receives the query from the user interface and analyzes it based on the mediated ontology for discovering the appropriate EHR DaaS. The second module receives the set of semantic descriptions of discovered EHR DaaSs and applies a query rewriting algorithm that generates a set of valid and executable of EHR DaaS compositions. The third one iteratively processes each rewriting previously generated in order to detect incompatibilities (semantic conflicts at the data level) and invokes the appropriate mediation services for their resolution. Mediation services offer the conversion functions defined in the conversion repository and referenced by SDO concepts. The last module arranges the selected EHR DaaSs along with the added mediation services in a composition plan which will be executed to return the results to users.

- **Interface Level:** The aim of this layer is to provide the interface for user whereby he can perform a query and receive results sets.

4.2 Models for Electronic Health Record Data-as-a-Services Composition

In this section, we propose models to address the issues related to query processing (query rewriting and conflict resolution) for EHR DaaSs composition. First, we formalize the notion of mediated ontology, with the introduction of generic and specific domain ontologies, which are useful as a support for semantic-aware querying and annotation of EHR DaaSs. Then, we propose a model for representing conjunctive queries over a mediated ontology. Finally, we develop models for EHR DaaS and mediation services.

Mediated Ontology

Mediated ontology includes two ontologies, namely, the generic and specific domain ontologies which have 'GDO' and 'SDO' as namespaces for their respective concepts. Such ontology should be defined by domain experts and specified using RDF/RDFS. The generic and specific ontologies models are inspired from [6] [31] [25]. In order to provide a precise semantic annotation for our service model we use these two models.

Definition 1. *(Generic Domain Ontology) : A RDFS generic ontology is 6-tuple $<$ C,D,OP, DP, SC, SP$>$ where*

- *C is a set of classes.*
- *D is a set of data types.*
- *OP is a set of object properties. Each object property has its own domain and range in C.*
- *DP is a set of data type properties. Each data type property has a domain in C and range in D.*
- *SC is a relation over $C \times C$, representing the sub-class relationship between classes. For example C_2 SC C_1 expresses that C_2 is subclass of C_1.*
- *SP is a relation over $(OP \times OP) \cup (DP \times DP)$, representing the sub-property relationship between homogeneous properties. For example DP_2 SP DP_1 means that DP_2 is a sub-property of DP_1.*

Figure 4 depicts the Generic Domain Ontology, in which class nodes are represented by ovals and data type nodes are represented by rectangles. In the GDO ontology, the GDO:Patient class is a core concept that characterizes patient information, such as name, SSN, etc. The GDO:Laboratory-test-order class captures information on laboratory tests ordered for a patient. It is related to the GDO:Patient class through the object property GDO:Has-order. The GDO:Laboratory-test-Result class captures results of laboratory tests ordered for patient and is related to the GDO:Laboratory-test-order class via the GDO:Has-Result object property. Individuals of GDO:Laboratory-test-result such as LDL, AST, ALT, TotalBilirubin, etc. are subclasses of the GDO:Panel class. The individuals that belong to the GDO:Laboratory-test-Result class may be related to multiple panels and each panel has several laboratory tests. The GDO:Disease class characterizes the disease which can be indicated by patient GDO:Laboratory-test-result and is related to the laboratory test class via the GDO:indicate object property.

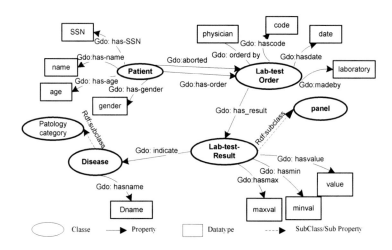

Fig. 4. Generic Domain ontology

Definition 2. *(Specific Domain Ontology)* : *A RDFS specific ontology is 3 tuple* $< C_g, C_i, \tau >$, *where:*

- C_g is a set of concepts that represent the different conflictual aspects of a generic concept in Generic Domain Ontology (GDO). Each C_g has a name and a set of specialized concepts (i.e. sub concepts); the name represents a conflictual aspect of the associated generic concept. In the example depicted in Figure 5, SDO:Laboratory-test-code and SDO:disease-code are C_g concepts.
- C_i is a distinct set of concepts having the same super-concept C_g. By definition, C_i are not allowed to have sub-concepts. The properties of C_i are defined as follows :
  - name of concept.
  - Seqno is the property that represents the sequence number of a C_i concept among its siblings.
  - A couple of properties reference the conversion functions between orderly organized object of C_i. The function name denotes the conversion from C_i to subsequent or precedent sibling, for instance snomed-to-loinc, loinc-to-snomed or mg/l-to-mmol/l, as it follows the mapping direction. Supported conversions between sibling subclasses are $n \longrightarrow 1$ and $1 \longrightarrow 1$.
- τ refers to the sibling relationships on C_i and C_g. The relationships among elements of C_g is disjoint. However elements of C_i of a given C_g have peer relationship. They have similar data semantics, so that conversion or mapping can be performed among them.

Let us illustrate this definition with an example in Figure 5. The concept GDO:Laboratory-test-Result in that figure has a conflictual aspect called "unit" that is described as a member of C_g in SDO (i.e. $SDO : unit$). The defined concept SDO:unit is linked to GDO:Laboratory-test-Result via the object property SDO:has-Unit which is also defined in SDO. SDO:unit has different measurement units represented as sub classes $C_i = \{mg/l, mmol/l, ..., n\}$. The code is also a conflictual aspect to both GDO:Laboratory-test-result and GDO:Disease concepts; i.e. codes can be represented differently in different health ontologies using $C_i = \{SNOMED, ICD, ..., n\}$. Note that, we can use an rdfs:collection to denote the sequence relationships between elements of C_i and typical processing will be to select one of the members of the container.

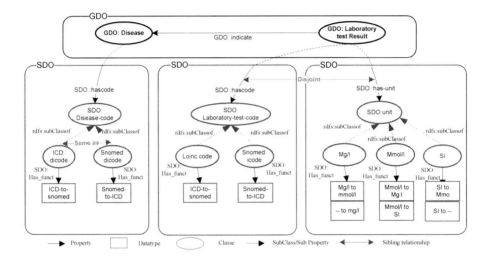

Fig. 5. Specific Domain ontology

Conjunctive Queries: In this paper we address conjunctive queries expressed using SPARQL, the do facto query language for the Semantic Web[9].

Definition 3. *A conjunctive queries Q has the form: $Q(X)$:-$< G(X,Y), C_q >$ where :*

- *$Q(X)$ is the head of Q, it has the form of relational predicate and represents the result of query.*
- *$G(X,Y)$ is the body of Q, it contains a set of RDF triples where each triple is of the form (subject. property.object). X and Y are called the distinguished and existential variables respectively. X and Y are subjects and objects in the RDF triples.*
- *$C_q = \{C1_q, C2_q, ..., Cn_q\}$ is a set of constraints expressed on X and Y variables in terms of traditional intervals or arithmetic expression like $x\theta constant$, $y\theta constant$ and where $\theta \in \{<,>\leq,\geq\}$.*

---

[9] SPARQL : http://www.w3.org/TR/rdf-sparql-query/

In our work, queries are formulated in SPARQL and use concepts from the mediated ontology (GDO ontology) and properties from the specific ontologies (SDO ontologies). Thus, a query can be seen as a graph with two types of nodes; class and literal nodes. Class nodes refer to classes in the ontology. They are linked via object properties. Literal nodes represent data types and are linked with class nodes via data type properties. Figure 6 depicts the RDF graph of the query Q_1 described in our scenario. The graph shows that Q1 has four class nodes P, LO, LR, D linked by object property GDO:has-order(P,LO), GDO:has-result(LO,LR) and GDO:indicate(LR,D) respectively.

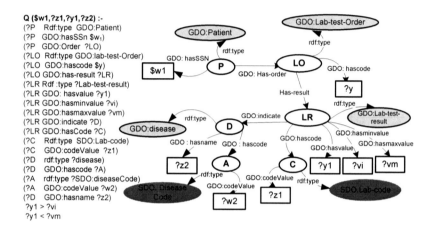

Fig. 6. Query in the running example

Electronic Health Record Data-as-a-Service model

We deem appropriate to follow the work of [6,34,10] to formalize the modeling of EHR DaaS as RDF views over a mediated ontology.

Definition 4. *EHR DaaS S_j is described as view in a Datalog-like notation over a GDO and SDO thus S_j model is :*

$$S_j(\$X_j, ?Y_j) :- < G_j(X_j, Y_j, Z_j), Co_j > |\alpha_{X_j}, \alpha_{Y_j} \text{ where:}$$

- X_j and Y_j are the sets of input and output variables of S_j, respectively,
- G_j represents the functionality of the EHR DaaS which is described as a semantic relationship between input and output variables.
- Z_j is the set of existential variables relating X_j and Y_j.
- $Co_j = \{Co_{j_1}, ..., Co_{j_n}\}$ is a set of constraints expressed on X_j, Y_j or Z_j variables like $x\theta constant$ and $y\theta constant$ where $\theta \in \{<, >\leq, \geq\}$.
- α_{X_j} and α_{Y_j}, named adornment, are a set of RDF triplets describing the semantic (ontological reference, unit...etc) or domain expression of X_j and Y_j respectively. Each adornment α is indicated by the 2-tuple; $< C_g, C_i >$ where : C_g : is an SDO concept that represent the different conflictual aspects X_j and Y_j); C_i : is a concept from SDO inherited from C_g .

An EHR DaaS model is described over a GDO and adorned by the entities from SDO. As an EHR DaaS is modeled uniquely over the entities of GDO, it does not provides explicit semantics about its input-and output parameters, so we extend its description with additional information describing more precisely how the semantics of the GDO concepts are described according to the SDO ontology. Then, each EHR DaaS model will be expressed as an adorned query [10]. The adornment is an annotation on variables, appearing in input and output parameters of a given EHR DaaS and expressed in term of SDO.

Figure 7 gives RDF view of EHR DaaS S_{21} and S_{22} services depicted in Table 1 with an adornment depicted in red color.

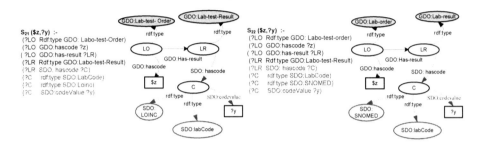

Fig. 7. Electronic Health Record Data-as-a-Service model

Mediation Service Model

Mediation Services are also represented as an EHR DaaS model (expressed in term of SDO ontology) whereas their adornments are described as a set of RDF triples that define the conversion function between peers of $SDO : C_i$ subconcepts from the same $SDO : C_g$ concept in a declarative way. We remind the reader that the different $SDO : C_i$ are organized as an ordered list, hence a conversion from one to another is always a concatenation of conversion functions.

Definition 5. *Mediation Service S_j is modeled as below :*
$S_j(\$I_j, ?O_j) : - < G_j(I_j, O_j, Z_j) > |\alpha_{Func<I_j,O_j>}$; *Where :*

- $\$I_j$ *defines the input parameter required for using mediation service;*
- $?O_j$ *defines the output parameter required for using mediation service;*
- Z_j *represents variables or constants generated inside a conversion or required during conversion.*
- $\alpha_{Func<I_j,O_j>}$ *represents the conversion function from $SDO : I_j$ to $SDO : O_j$.*

Figure 8 illustrates the RDF view of a mediation DaaS service $S_{LOINC-SNOMED}$ utilized for converting a labcode from S_{21} to S_3.

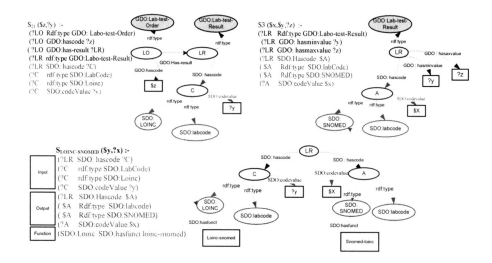

Fig. 8. Mediation Service model

5 Query Processing for Electronic Health Record Data-as-a-Services Composition

In this section, we outline the development of query processing for EHR DaaSs composition and we give a detailed description of the key phases of query rewriting and conflict detection and resolution phases.

5.1 Query Processing Phases

The complete query processing steps are depicted in Figure 9. They include four processes. First, query formulation and service discovery, second, query rewriting, third, conflict detection and resolution, and finally query execution and result restitution.

1. Firstly, query formulation and service discovery: In (1) and (2) the user issues SPARQL queries in terms of mediated ontology. Doing so, in (3) and (4) the service selector discovers EHR DaaSs from the service registry that partially or completely matches the query entities (class nodes, object property nodes).
2. Secondly, query rewriting: In (5), the query sent over the mediated ontology is rewriten into a query that refers directly to the set of discovered EHR DaaSs. The query rewriting module of EHR DaaSs composition component uses an approach in the spirit of the "bucket algorithm" [18] that returns the maximally contained rewritings of a query. The algorithm computes for each query entity (class node, object property node) called bucket or subgoal in the bucket algorithm, the EHR DaaS that are relevant to it. Thus,

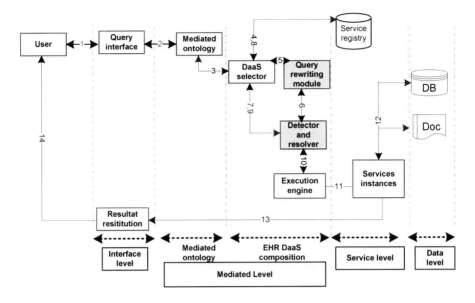

Fig. 9. Electronic Health Record Data-as-a-Services composition process

the rewriting is constructed by combining one element of every bucket. Candidate solutions generated by the query rewriting algorithm must be tested by applying the conjunctive query containment to validate it.
3. Thirdly, conflict detection and resolution: In (6), considering each combination generated by the query rewriting module, which may encompasses semantic conflicts at the data level (7), requires testing any conflict by comparing output and input of subsequent EHR DaaS in each query rewritings. The conflict is resolved with the insertion of a call to mediation services (8,9). Thus, besides EHR DaaSs, in each query rewriting combination mediation services are added to resolve conflicts.
4. Fourthly, query execution and result restitution: In (10, 11, 12), orchestrating the conflict-free composite service that has been generated requires a translation into an execution plan describing the data and control flows. Finally, (in 13 and 14) the result restitution module synthesizes results and returns them to users through user interface.

5.2 Query Rewriting

Given a query Q and a set of EHR DaaSs, the query rewriting module rewrites Q as composition of EHR DaaSs whose union of RDF graphs covers the RDF graph of Q. The query rewriting phase is preceded by a preprocessing step (carried out prior to receiving the composition query) in which the RDF views are extended with the RDFS semantic constraints (i.e. *subClassOf*, *subPropertyOf*, *Domain* and *Range*) to obtain a better matching with the composition queries.

Our composition query rewriting algorithm [6] has two main phases detailed as follows:

(1) **Finding the Covered Query's Sub-graphs:** In this phase the query is compared to the different RDF views to determine the class nodes and the object properties that are covered by the views. The term covers means that there is a containment mapping between classes nodes and object properties of Q and those of the views. The result of this step is a table summarizing for each EHR DaaS, the covered classes nodes and object properties. For example, the table2 shows the covered classes and properties for the services S_{11}, S_{21}, S_3, S_4.

Table 2. The covered classes nodes and object properties for the services S_{11}, S_{21}, S_3, S_4

Service	Covered classes and properties
$S_{11}(\$w_1, ?y)$	Patient(w_1), has-order(Patient, LabOrder), LabOrder(y)
$S_{21}(\$y, ?y_1, ?z_1)$	LabOrder(y), hasResult(LabOrder, LabResult), LabResult(y_1), has-Code(LabResult,LabCode), LabCode(z_1)
$S_3(\$z_1, ?v_i, ?v_m)$	LabResult(v_i, v_m), hasCode(LabResult, LabCode), LabCode(z_1)
$S_4(\$z_1, ?z_2, ?w_2)$	LabCode(z_1), hasCode(LabResult, LabCode), indicate (LabResult, Disease), Disease(z_2), hasCode(Disease, DiseaseCode), DiseaseCode(w_2)

(2) **Composition Generation:** In the second phase, the query rewriting module combines the different lines in the generated table in such a way all the classes nodes and object properties of the query are covered by the combination and the combination is executable. A combination is executable if the inputs of involved services are bound or can be made bound by other services (whose inputs are bound). For example, the combination of the services S_{11}, S_{21}, S_3, S_4 covers the whole set of classes nodes and properties; all the inputs of theses services are bound or can be made bound. Therefore the set S_{11}, S_{21}, S_3, S_4 is considered as a valid composition.(see Figure 10)

Fig. 10. Composition generation

5.3 Conflicts Detection and Resolution

In this phase we detect and resolve the semantics conflicts in the EHR DaaS compositions generated in the previous phase. This phase includes the following steps:

(1) The localization of conflicts between interconnected EHR DaaSs. Conflicts arise when data elements that have to be exchanged between two interconnected EHR DaaSs are interpreted differently by these services. Conflicts localization is accomplished by the *conflict detection module*. This kind of conflict is named *Attribute level incompatibilities* by the classification of structural and semantic message level heterogeneities proposed in [29]. According to that work, attribute level incompatibilities arise when semantically similar attributes are modeled using different descriptions. To detect the conflicts we define some rules that define some conflicts between EHR DaaSs expressed in RDF as indicated bellow:

– let $SDO : R_i$ and $SDO : E_i$ be subclasse of the same conflictual class $SDO : C_g$, such as $SDO : lab - code$, thus :
```
SDO:Ei   rdfs:subClassof   SDO:Cg
SDO:Ri   rdfs:subClassof   SDO:Cg
```
– Then, if we have two EHR DaaS S_i and S_j including concepts $SDO : E_i$ and $SDO : R_i$ respectively in their RDF descriptions as an adornment, expressed as depicted by the following triples :
```
LR   SDO:hascode   ?N
?N   rdf:type      SDO:Ei
?N   rdf:type      SDO:LabCode
LR   SDO:hascode   ?A
?A   rdf:type      SDO:Ri
?A   rdf:type      SDO:labCode
```
then we have code laboratory test conflict.

In sum, the set of conflict types identified in our solution is the set of conflictual concepts $SDO : C_g$. For instance, code disease conflict or unit conflict. Other conflict types (e.g. data representation, data precision ,... etc) can be added to SDO in order to resolve more semantic conflicts at the data level. As a consequence, this step will provide the set of conflict objects where each conflict object will be identified as 3 tuple $< O(S_i), I(S_j), C_g >$ where $O(S_i)$ is an adorned output parameter of a given EHR DaaS source S_i , $I(S_j)$ is an adorned input parameter of a given EHR DaaS target S_j, conflict type is a member of set of conflictual concepts $C_g = \{unit, labCode, ..., n\}$.

To detect the conflicts, the algorithm depicted (algorithm 1) will take each composition (represented as Directed Acyclic Graph) and iteratively verify the rules expressed previously for each parameter (adornment only) exchanged between interconnected services to find out all possible conflicts which will be stored in the conflicts-objects set.

(2) The conflict objects detected previously will be resolved by the automatic invocation of an appropriate mediation service. The latter is identified through:
 – the input parameter $SDO : O(S_i)$, which is the output of S_i;
 – the output parameter $SDO : I(S_j)$ which is the input of S_j;
 – the conversion function as an adornment, defined as property of $SDO : O(S_i)$ and targets $SDO : I(S_j)$.

Algorithm 1. semantic conflict detection and resolution

Require: $M_{i,j}$ {Matrix is a graph of EHR DaaS combination with conflict}; $i,j,k,z \in \mathbb{N}$, $\mathcal{CO}$ a set of Conflict Object,
1: {Detection conflict step}
2: **for** $i = 1$ to n **do**
3:   **for** $j + 1$ to n **do**
4:     **if** $M[i][j] = 1$ **then**
5:       **if** $Output.S_i$ AND $Input.S_j$ have the same conflictual concept as type and differents SDO subclasses **then**
6:         $\mathcal{CO}z$ = New conflict object($output.S_i$, $input.S_j$, conflictual concept),
7:         Add ($\mathcal{CO}$, $\mathcal{CO}z$)
8:       **end if**
9:     **end if**
10:   **end for**
11: **end for**
12: {Resolution conflict step}
13: **for each** $\mathcal{CO}z$ in $\mathcal{CO}$ **do**
14:   {according to conflict object identify mapping function $(Output.S_i, Input.S_j)$ from SDO ontology}
15:   $M[i,j] = 0$ {delete S_i and S_j arc}
16:   ADD S_K {ADD mediation DaaS service}
17: **end for**
Ensure: $M'_{i,j}$ graph of EHR DaaS combination without conflict

As a consequence of this phase, the mediation services $S_{LOINC-SNOMED}$ and $S_{mmol/l-mg/l}$ are added to the first and second EHR DaaS compositions to resolve conflict as depicted in figure (see Figure 11). Afterwards, the obtained conflict-free compositions will be translated into execution plans (i.e. orchestrations) describing the data and control flows as depicted in the same Figure 11. For space limitation, we do not detail this step in the paper.

6 Implementation and Evaluation

To illustrate the viability of our approach to EHR DaaS composition, we implemented about /411/ EHR DaaS Web services on top of a set of medical data sources containing synthetic data about patients, including information like diseases, medical tests, allergies, medications lists, vaccination records, ongoing treatments, consultations, personal information (e.g., date of birth, sex, etc), etc. All of these data are usually represented by the commonly used types of the EHR information model. We built a medical ontology based on the building blocks and the data-types defined in the HL7 and the openEHR standards. The ontology included /81/ ontological concepts and /413/ properties (i.e., both datatype and object properties). We modeled all services as RDF views over that ontology. These views were used to annotate the description files (WSDLs) of corresponding DaaS services. We implemented also a set of mediation services; these services were used to convert the values of exchanged data from HL7 to

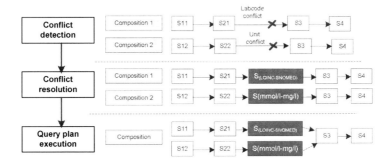

Fig. 11. Conflict detection and resolution

openEHR specific datatypes (and vice versa). These services allowed for example to convert between different medical data's measurements units, precisions, etc. All services were implemented in Java, and hosted on a GlassFish Web server.

Figure 12 depicts our implemented evaluation system. In that figure, the *DaaS Web services layer* plays the role of an abstraction layer on top of heterogeneous medical data sources; i.e., medical data located in heterogeneous data sources (e.g., relational data bases, silos of data-centric homegrown or packaged applications (e.g. SAP, PeopleSoft, Siebel, etc), files XML files, etc.) are all accessed by the same interface, the Web service interface. These services can be composed by the *Web Service Management System*. The system's users are assisted in formulating their queries (i.e., SPARQL queries) over the ontology.

We tested our system with a set of real-life queries (including that of the running example), examples included: "Q_1: check whether the medication ABC identified by the code "801" to be prescribed to patient John Doe interacts with the ones currently taken by that patient", "Q_2: For any given social security number X of a patient and a medication code Y representing the medication to be prescribed, verify whether the medications taken by the patient may interact with Y", "Q_3: What are the tests performed by patients that have been administered a given medication?", etc.

Throughout our tests, we made the following observations: (i) the system was able to process hundreds of services in a reasonable time (411 services in less than one second), the reported time is the time to create the composite services (this involves parsing the WSDL files of services, determining the relevant ones and building the composition); the created compositions included both the EHR DaaS services and the necessary mediation services. (ii) in all of the considered queries (20 queries), the system was able to insert the necessary data mediation services to transform data between heterogeneous component services in a composition. The system users (mainly physicians in our tested examples) have expressed their satisfaction, as they were able to answer their queries on the fly without any programming involved. Users used the created compositions in their daily clinical scenarios (e.g., prescribing a medication, studding the risks of a medication, etc.).

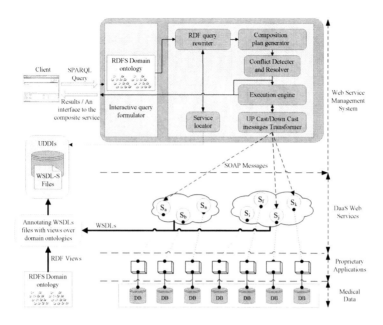

Fig. 12. The implemented system for evaluating our approach

7 Related Works

In this section, we give an overview of the main research works related to our subject. We have categorized these works into the following areas: EHR data integration, EHR Web service composition and query rewriting approach for automatic Web service composition.

7.1 Electronic Health Record Data Integration

Data integration is the problem of combining data residing at different sources to provide the user with a unified view of data. There is a large body of research work on data integration not only in the area of computer science but also in health and biomedical informatics[1] [11]. Broadly speaking data integration approaches can be classified into two main approaches: "data warehouses" and "mediation" approaches.

The mediation approach, unlike data warehouse, does not materialize data; it integrates data at the query processing time. In the e-health area most data integration projects have adopted this approach like, to name a few, Synapses [16], Synex [16] and Pangea-LE [1], etc. Most of these projects provide a global XML schema over structured XML views of EHR documents stored for a patient in existing health data repositories. The mediated system accepts requests for data from clients, decomposes them into queries against the connected data sources, and integrates the responses dynamically. In contrast with these projects, we adopt a service-oriented data mediation architecture and a service composition approach to integrate data on the fly by composing autonomous EHR DaaSs. Also, unlikely to these projects we handle the semantic conflicts at the data value level when data is exchanged among healthcare systems.

7.2 Medical Web Services Composition

A large number of research work have addressed the problem of WS composition in the healthcare application domain[2], [28], [39] and [21]. However, the bulk of these works have focused only on workflow oriented compositions; i.e. compositions that implement the different steps involved in a given business task (e.g. patient admission in a hospital, patient discharge, etc). We review some of these works in the following:

Authors in [2] defined a model-driven approach for semi-automatic Web service orchestration with run-time binding in the healthcare domain. Information related to medical Web services can be found in the corresponding standardization documents for instance HL7, DICOM and IHE. Unfortunately, WS composition in that work does not allow to integrate heterogeneous medical data sources.

Budgen et al. [9] propose data Integration Broker for Heterogeneous Information Sources (IBHIS). The proposed broker adopts a service-based model to query data at autonomous healthcare agencies. The broker achieves its goals through the use of semantic data descriptions, a semantic registry and a query engine. The semantic registry is based upon an extended form of UDDI, incorporating a matchmaker to match OWL-S data descriptions from the registry to the users' queries. The query engine formulates the users' queries, interacts with the matchmaker to answer the query, and displays the final results to users. Unlink to our work, in that work there is no way to compose services to address user's complex queries (i.e. the work assumes that a query can be always resolved by one service). Furthermore, the use of OWL-S language alone does not allow to specify explicitly the relationship between inputs and outputs of a DaaS service which may lead to errors in the service matching phase. Further, IBHIS relies on SNOMED as a mediated ontology which is a restriction, as in our vision we should remain independent from a specific ontology.

Hristoskova et al. [20] present an implementation of a dynamic and automatic composer for medical support services in the ICU (The Intensive Care Unit). The composition is achieved by semantically described Web services in order to provide automatic WS-BPEL composition. In comparison with our work, this solution does not address the EHR data integration using WS composition. Also, it is restricted to intensive care unit in hospital where data heterogeneity is not a real concern.

In ARTEMIS project [12] [8] ensures the semantic interoperability of Electronic Health Records and medical applications through using the Web service technology. It presents a mechanism for publishing, discovering and invoking semantically enriched Web services in Peer-to-Peer medical data sharing environments. Web services are annotated with OWL medical ontologies. However, the ARTEMIS project does not provide means to compose individual medical data services to answer the user's complex queries. Furthermore, heterogeneities at the data value level was not addressed in that work.

7.3 Query Rewriting Approach for Web Service Composition

Automatic Web service composition approaches can be classified according to the techniques adopted to solve the composition problem into: AI planning based approaches, workflow based approaches and query rewriting based approaches [6,40,27,5,36,38,32]. We review in the following some works in the last category as they relate to our approach.

Lu et al. [27] provide a framework for answering queries with a conjunctive plan that includes input and outputs of participating Web services annotated with Datalog expressions. In [36], a combination of inverse rules query reformulation algorithm and tuple filtering are used to generate a universal integration plan or a composition to answer user queries. However, those works do not take into account the semantics of the services during the matching phase; i.e. services are matched based on types matching. In order to address this problem, Bao et al. [5] proposed a semantic query rewriting approach using the CARIN [15] language. However, unlike to our work the semantic query rewriting in that work is limited to one-to-one matching; i.e. the work assumes that a query can be resolved by one service and there is now need to service combination.

Also, Zhou et al. [40] introduce an ontology-based approach to publishing and composing data-intensive Web services. They propose an extension to the description capability of OWL-S. This extension has the form of a SPARQL query defining semantic content and constraints on data published by the service. Also, an algorithm to generating service composition based on ontology language and graph-based planning are outlined. However, our work relies on a more efficient RDF query rewriting algorithm [6] that uses many optimization heuristics to out speed their proposed algorithm. Furthermore, we address data values heterogeneities in the obtained compositions.

Furthermore, Vaculin et al. [38] describes mechanisms for specification of generic data providing services using RDF views. They provided a characterization of matching conditions for DaaS services and developed an algorithm for matching with calculation of a matching degree between service requests and service advertisements. However, that work overlooks the relationships between matched outputs and inputs in the RDF graph of the DaaS service which may lead to erroneous matching results.

8 Perspectives and Conclusions

In this paper, we proposed an approach to automatically compose EHR DaaS Web services published by heterogeneous health information systems that employ different EHR data standards. The proposed approach follows a local-as-view paradigm by explicitly requiring a two-level mediated ontology. The first level models the generic data concepts and their inter-relationships while the second allows specifying in a declarative way how a concept of the first level is represented in different health ontologies and data standards. These ontologies are utilized to annotate EHR DaaS and mediation services on the one hand, and to specify user queries on the other hand. User queries are rewritten in terms of EHR DaaS services using an efficient query rewriting algorithm. Furthermore, our approach makes use of mediation services to handle the semantic heterogeneity of exchanged data.

As future work, we intend to improve our mediation approach in order to address structural-level incompatibilities, as well as complex data transformations between input and output parameters of EHR DaaSs. In order to realize this objective, we plan to study different solutions for the composition of mediation services to ensure a complete mediation in EHRs DaaS composition. In addition, we intend to include that data quality aspects (e.g. data trustworthiness and provenance, etc) of data provided by the EHR DaaSs in our composition approach. We will investigate the use of RDF reification and named graphs in that respect.

References

1. Angulo, C., Crespo, P., Maldonado, J.A., Moner, D., Pérez, D., Abad, I., Mandingorra, J., Robles, M.: Non-invasive lightweight integration engine for building ehr from autonomous distributed systems. International Journal of Medical Informatics 76(suppl.3), 417–424 (2007)
2. Anzböck, R., Dustdar, S.: Modeling and implementing medical web services. Data & Knowledge Engineering 55(2), 203–236 (2005)
3. Arguello, M., Des, J., Perez, R., Fernandez-Prieto, M., Paniagua, H.: Electronic health records (ehrs) standards and the semantic edge: A case study of visualising clinical information from ehrs. In: International Conference on Computer Modeling and Simulation, pp. 485–490 (2009)
4. Austin, T., Kalra, D., Tapuria, A., Lea, N., Ingram, D.: Implementation of a query interface for a generic record server. I. J. Medical Informatics 77(11), 754–764 (2008)
5. Bao, S., Zhang, L., Lin, C., Yu, Y.: A semantic rewriting approach to automatic information providing web service composition. In: Mizoguchi, R., Shi, Z.-Z., Giunchiglia, F. (eds.) ASWC 2006. LNCS, vol. 4185, pp. 488–500. Springer, Heidelberg (2006)
6. Barhamgi, M., Benslimane, D., Medjahed, B.: A query rewriting approach for web service composition. IEEE Transactions on Services Computing 3, 206–222 (2010)
7. Beale, T.: Archetypes: Constraint-based Domain Models for Futureproof Information Systems
8. Bicer, V., Laleci, G.B., Dogac, A., Kabak, Y.: Artemis message exchange framework: semantic interoperability of exchanged messages in the healthcare domain. SIGMOD Rec. 34, 71–76 (2005)
9. Budgen, D., Rigby, M., Brereton, P., Turner, M.: A data integration broker for healthcare systems. Computer 40, 34–41 (2007)

10. Calvanese, D., Giacomo, G.D., Lenzerini, M., Nardi, D., Rosati, R.: A principled approach to data integration and reconciliation in data warehousing. In: Proceedings of the International Workshop on Design and Management of Data Warehouses, DMDW 1999 (1999)
11. Cruz-Correia, R., Vieira-Marques, P., Ferreira, A., Almeida, F., Wyatt, J., Costa-Pereira, A.: Reviewing the integration of patient data: how systems are evolving in practice to meet patient needs. BMC Medical Informatics and Decision Making 7(1), 14 (2007)
12. Dogac, A., Laleci, G.B., Kirbas, S., Kabak, Y., Sinir, S.S., Yildiz, A., Gurcan, Y.: Artemis: Deploying semantically enriched web services in the healthcare domain. Information Systems 31(4-5), 321–339 (2006); The Semantic Web and Web Services
13. Dolin, R., Alschuler, L., Beebe, C., Biron, P., Boyer, S., Essin, D., Kimber, E., Lincoln, T., Mattison, J.: The HL7 clinical document architecture. Journal of the American Medical Informatics Association 8(6), 552 (2001)
14. Eichelberg, M., Aden, T., Riesmeier, J., Dogac, A., Laleci, G.B.: A survey and analysis of electronic healthcare record standards. ACM Comput. Surv. 37(4), 277–315 (2005)
15. Goasdoué, F., Lattès, V., Rousset, M.-C.: The use of carin language and algorithms for information integration: The picsel system. Int. J. Cooperative Inf. Syst. 9(4), 383–401 (2000)
16. Grimson, J., Stephens, G., Jung, B., Grimson, W., Berry, D., Pardon, S.: Sharing health-care records over the internet. IEEE Internet Computing 5, 49–58 (2001)
17. Halevy, A.Y.: Theory of answering queries using views. SIGMOD Rec. 29(4), 40–47 (2000)
18. Halevy, A.Y.: Answering queries using views: A survey. The VLDB Journal 10, 270–294 (2001)
19. Hori, M., Ohashi, M.: Applying xml web services into health care management. In: Hawaii International Conference on System Sciences, vol. 6, p. 155a (2005)
20. Hristoskova, A., Moeyersoon, D., Hoecke, S.V., Verstichel, S., Decruyenaere, J., Turck, F.D.: Dynamic composition of medical support services in the icu: Platform and algorithm design details. Computer Methods and Programs in Biomedicine 100(3), 248–264 (2010)
21. Kart, F., Miao, G., Moser, L., Melliar-Smith, P.: A distributed e-healthcare system based on the service oriented architecture. In: IEEE International Conference on Services Computing, SCC 2007, pp. 652–659 (2007)
22. Katehakis, D.G., Sfakianakis, S., Kavlentakis, G., Anthoulakis, D.N., Tsiknakis, M.: Delivering a lifelong integrated electronic health record based on a service oriented architecture. IEEE Transactions on Information Technology in Biomedicine 11(6), 639–650 (2007)
23. Lee, Y., Patel, C., Chun, S.A., Geller, J.: Towards intelligent web services for automating medical service composition. In: Proceedings of the IEEE International Conference on Web Services, ICWS 2004, p. 384. IEEE Computer Society Press, Washington, DC, USA (2004)
24. Lenz, R., Beyer, M., Kuhn, K.A.: Semantic integration in healthcare networks. International Journal of Medical Informatics 76(2-3), 201–207 (2005); Connecting Medical Informatics and Bio-Informatics - MIE 2005
25. Liu, Q., Huang, T., Liu, S.-H., Zhong, H.: An ontology-based approach for semantic conflict resolution in database integration. Journal of Computer Science and Technology 22, 218–227 (2007), doi:10.1007/s11390-007-9028-4
26. Liu, S., Ni, Y., Mei, J., Li, H., Xie, G., Hu, G., Liu, H., Hou, X., Pan, Y.: ismart: Ontology-based semantic query of cda documents

27. Lu, J., Yu, Y., Mylopoulos, J.: A lightweight approach to semantic web service synthesis. In: WIRI 2005, Proceedings of International Workshop on Challenges in Web Information Retrieval and Integration, pp. 240–247 (2005)
28. Mykkänen, J., Riekkinen, A., Sormunen, M., Karhunen, H., Laitinen, P.: Designing web services in health information systems: From process to application level. International Journal of Medical Informatics 76(2-3), 89–95 (2005)
29. Nagarajan, M., Verma, K., Sheth, A., Miller, J.: Ontology driven data mediation in web services. International Journal of Web Services Research 4(4), 104–126 (2007)
30. Orgun, B., Vu, J.: Hl7 ontology and mobile agents for interoperability in heterogeneous medical information systems. Computers in Biology and Medicine 36(7-8), 817–836 (2006); Special Issue on Medical Ontologies
31. Ram, S., Park, J.: Semantic conflict resolution ontology (scrol): an ontology for detecting and resolving data and schema-level semantic conflicts. IEEE Transactions on Knowledge and Data Engineering 16(2), 189–202 (2004)
32. Rao, J., Su, X.: A survey of automated web service composition methods. In: Cardoso, J., Sheth, A.P. (eds.) SWSWPC 2004. LNCS, vol. 3387, pp. 43–54. Springer, Heidelberg (2005)
33. Sachdeva, S., Bhalla, S.: Semantic Interoperability in Healthcare Information for EHR Databases. In: Kikuchi, S., Sachdeva, S., Bhalla, S. (eds.) DNIS 2010. LNCS, vol. 5999, pp. 157–173. Springer, Heidelberg (2010)
34. Saleh, I., Kulczycki, G., Blake, M.B.: Demystifying data-centric web services. IEEE Internet Computing 13, 86–90 (2009)
35. Stroetmann, K., Stroetmann, V.: Towards an Interoperability Framework for a European e-Health Research Area–Locating the Semantic Interoperability Domain. In: EC Workshop on semantic interoperability, Brussels, pp. 14–15 (February 2005)
36. Thakkar, S., Ambite, J.L., Knoblock, C.A.: A data integration approach to automatically composing and optimizing web services. In: Proceedings of the ICAPS Workshop on Planning and Scheduling for Web and Grid Services (2004)
37. Truong, H.L., Dustdar, S.: On analyzing and specifying concerns for data as a service. In: APSCC, pp. 87–94 (2009)
38. Vaculín, R., Chen, H., Neruda, R., Sycara, K.: Modeling and discovery of data providing services. In: Proceedings of the 2008 IEEE International Conference on Web Services, pp. 54–61. IEEE Computer Society, Washington, DC, USA (2008)
39. Wright, A., Sittig, D.F.: Sands: A service-oriented architecture for clinical decision support in a national health information network. J. of Biomedical Informatics 41, 962–981 (2008)
40. Zhou, L., Chen, H., Wang, J., Zhang, Y.: Semantic web-based data service discovery and composition. In: International Conference on Semantics, Knowledge and Grid, pp. 213–219 (2008)

A Modular Database Architecture Enabled to Comparative Sequence Analysis

Paola Bonfante[1], Francesca Cordero[2,3], Stefano Ghignone[1], Dino Ienco[2], Luisa Lanfranco[1], Giorgio Leonardi[4], Rosa Meo[2], Stefania Montani[4], Luca Roversi[2], and Alessia Visconti[2]

[1] Dipartimento di Biologia Vegetale, Università di Torino, Italy
[2] Dipartimento di Informatica, Università di Torino, Italy
[3] Dipartimento di Scienze Cliniche e Biologiche, Università di Torino, Italy
[4] Dipartimento di Informatica, Università del Piemonte Orientale, Italy
{fcordero,ienco,meo,roversi,visconti}@di.unito.it,
{giorgio.leonardi}@mfn.unipmn.it,
{stefania.montani}@unipmn.it,
{pbonfant,sghignon,llanfran}@unito.it

Abstract. The beginning of post-genomic era is characterized by a rising numbers of public collected genomes. The evolutionary relationship among these genomes may be caught by means of the comparative analysis of sequences, in order to identify both homologous and non-coding functional elements. In this paper we report on the on-going **BIOBITS** project. It is focused on studies concerning the bacterial endosymbionts, since they offer an excellent model to investigate important biological events, such as organelle evolution, genome reduction, and transfer of genetic information among host lineages. The **BIOBITS** goal is two-side: on the one hand, it pursues a logical data representation of genomic and proteomic components. On the other hand, it aims at the development of software modules allowing the user to retrieve and analyze data in a flexible way.

1 Introduction

Genomics and post-genomics studies which have bloomed in the last decade are offering new tools for applied biotechnological research in several fields from medical, pharmaceutical to industrial and environmental. Sequencing of the human genome has generated a great deal of interest in the diagnosis and treatment of diseases using genomic medicines. Structural genomics approaches covering topologically similar proteins or gene families are great assets for progress in the development of novel therapeutics. In addition the genomic analysis of microbial communities in a culture-independent manner (metagenomics) has also given the opportunity to probe and exploit the enormous resource represented by still underscribed microbial diversity.

This paper is an extension of a work already published [12]. It describes the on-going project BIOBITS[1] that aims at performing an extensive comparative genomic studies in order to answer fundamental questions concerning the biology, ecology and evolutionary history. The specific goal of BIOBITS is to get insights on the tri-partite system, constituted by (i) a bacterial endosymbiont of an arbuscular mycorrhizal (AM) fungus, (ii) AM fungi living in plant roots, and (iii) plant roots.

Bacterial endosymbionts are widespread in the animal kingdom, where they offer excellent models for investigating important biological events such as organelle evolution, genome reduction, and transfer of genetic information among host lineages [30]. By contrast, examples of endobacteria living in fungi are limited [26] and those best investigated live in the cytoplasm of AM fungi [9]. AM fungi are themselves obligate symbionts since, to complete their life cycle, they must enter in association with the root of land plants.

AM species belonging to the Gigasporaceae family harbour an homogeneus population of endobacteria which have been recently grouped into a new taxon named *Candidatus* Glomeribacter gigasporarum [7]. The AM fungus and its endobacterium Ca. Glomeribacter gigasporarum are currently used as a model system to investigate endobacteria-AM fungi interactions.

The project takes advantages by the employment of a massive large-scale analysis and genomic comparison study of phylogenetically related free-living bacteria. Moreover, the comparison with genomes of other endosymbionts species will provide insights about the reason of the strict endosymbiotic life-style of this bacterium.

Another aspect taken into account is the analysis of metabolic pathways. A strong reason of interest in this project is based on the assumption that the symbiotic consortia may lead to the discovery of molecules of interest for the development of novel therapies and other applications in biotech.

In this paper we report specifically on a step of BIOBITS whose goal, roughly, is the development of a modular database which allows to import, to store, and to analyze massive genomic data. Later in BIOBITS we will extensively develop a computational genomic comparison focused on the above bacterium and fungi genomes. BIOBITS deploys a data warehouse that stores in a multi-dimensional model the interesting components of the project. Such a component should have the following characteristics: i) being able to store genomic data from multiple organisms, possibly taken from different public database sources; ii) annotating the genomic data making use of the alignment between the given sequences and the genomic sequences of other similar organisms; iii) annotating the genomic sequences and the protein transcript products by the full use of ontologies developed by the biology and bioinformatics communities; iv) comparing and visually presenting the results of the genomic alignment; iv) being able to cluster genomic or proteomic data coming from different organisms. The aim is at finding easily

---

[1] BIOBITS is a project funded by Regione Piemonte under the Converging Technologies Call. BIOBITS involves Università di Torino, Università del Piemonte Orientale, CNR and the companies ISAGRO Ricerca s.r.l., GEOL Sas, Etica s.r.l.

increasing levels of similarity and induce on one side the steps of the phylogenetic evolution and on the other side investigate on the metabolic pathways.

As a matter of fact, we wish to take advantage of the possibilities offered by computer science technology and its methodologies to analyse the genomic data the project will produce. The analysis of genomic data requires computational tools that allow to "navigate" flexibly data from arbitrary (at least in principle) user defined perspectives and under different degrees of approximation.

In this paper we describe the BIOBITS system architecture in terms of BIO-BITS Data Mart and BIOBITS modules. With respect to the previous publication [12], we report a detailed description of two modules, namely *Case Base Reasoning* and *Co-clustering* modules, that have been developed to perform a comparative genomic analysis. Moreover, we show the results obtained in a case study by the use of the system. The case study shows the utility and flexibility of an integrated system whose modules allow to retrieve and analyze different portions of data, at the granularity level that is needed by the user. This flexibility eliminates the necessity of perform any pre-processing to the data in order to adapt it to the analysis algorithm and to the user's goal.

In the presented case study we extracted a set of biological sequences belonging to the organism under investigation by following the BIOBITS Data Mart star schema. BIOBITS project focus on the identification of the evolutionary relationships among species more similar to *Ca. G. gigasporarum*. Using the Case Base Reasoning module, we retrieved sequences that are similar to the given organism. Retrieval is performed according to the suitable abstraction level over the data given by a taxonomy of granularities. Finally, to the resulting sequences we applied the Co-clustering module and we were able to identify protein domains common among the sequences.

2 Related Works

There is a wide variety of approaches in designing tools to analyze biological data. Experience suggests that the best way to data analysis is to set up a database. An 'historical' example is ACeDB (*A C. elegans Database* [1]), one of the first hierarchical, rather than relational, model organism databases. Another example is ArkDB [21], a schema that was created to serve the needs for the subset of the model organism community interested in agriculturally important animals. ArkDB has been successfully used across different species by different communities, but is rarely used outside the agricultural community.

On "top" of databases a great variety of applications is available, from those ones for the annotation community to molecular pathway visualization, or from the work-flow management to the comparative genome visualization.

Currently, there is a rich community and many available software tools built around MAGE [27] and GMOD [33]. GMOD stands for Generic Model Organism Database project, which brought to the development of a whole collection of software tools for creating and managing genome-scale biological databases, in the forthcoming description. In the BIOBITS project GMOD and its database Chado have been selected as the data elaboration and management center.

2.1 GMOD and Chado Database

The BIOBITS software architecture is built upon a layer provided by GMOD system. We report here the main motivations that lead to this choice.

The design and implementation of database applications is time consuming and labor-intensive. When database applications are constructed to work with a particular schema, changes to the database schema require in turn changes to the software. Unfortunately, these changes are frequent in real projects due to changes in requirements. In particular they are frequent in bioinformatics. Most critical are the changes in the nature of the underlying data, which follow the current understanding of the natural world. Additional requirements are placed by the rapid technological changes in experimental methods and materials. Finally, the wide variety of biological properties in the organisms species always has made difficult to create a unique model schema valid for all the species.

All the above outlined motivations led to the design of Chado database model which is a generic and extensible model, whose software is available under an open source delivery policy. Chado schema can be employed as the core schema of any model organism data repository. This common schema increases interoperability between software modules that operate on it.

Chado data population is driven by *ontologies*, i. e. *controlled vocabularies*. Ontologies give a typing to the entities with the result of partitioning the whole schema into *subschemas*, called *modules*. Each *module* encapsulates a different biological domain and uses an appropriate ontology. An ontology characterizes the different types of entities that exist in a world under consideration by means of primitive relations. These primitives are easy to understand and to use, they are expressive and consistent, and they allow the reasoning about the concepts under representation. Typical examples of ontological relations are: (i) *is_a* which expresses when a class of entities is a subclass of another class, and (ii) *part_of* which expresses when a component constitutes a composite. Many other relation types are discussed in [15].

Concerning the schema of Chado it is worth remarking *feature* and *sequence* entities. *feature* allows both data and meta data; it can be populated by instances each determining the type of every other instance in the schema, in accordance with the ontology SO [15]. *sequence* contains biological sequence features, that include genetically encoded entities like genes, their products, exons, regulatory regions, etc... *feature* and *sequence* are further described by properties.

3 BIOBITS System Architecture

Here we deepen the description of the system which is designed to manage all the information and all the in-silico activities in the context of the project BIOBITS. This system is implemented through a modular architecture, described in detail in Section 3.2. The system architecture permits (1) to store and access locally all the information regarding the organisms to be studied, and (2) to provide algorithms and user interfaces to support the researchers' activities like: (i) searching and retrieving genomes, (ii) comparing and aligning with a genome of reference, (iii) investigating syntenies, and (iv) locally storing potentially new annotations.

The system architecture has been engineered exploiting the standard modules and interfaces offered by the GMOD project [33], and completed with custom modules to provide new functionalities. The main module of the system contains the database which provides all the data needed to perform the in-silico activities related to the project.

Thanks to the adoption of Chado database schema, on the one hand, we take advantage of its support in controlled vocabularies and ontologies. On the other hand, Chado is the standard database for most of the GMOD modules; therefore we can reuse these modules to support the main activities of the project and extend the system incrementally as the researchers' needs evolve. An example, is the possibility to use BioMart Chado's module which helps the user to identify the relevant dimensions of the problem, their hierarchies and to transform and import input data in the data warehouse conforming them in a typical star schema.

3.1 Star Schema in BIOBITS Data Mart

Essential in the data warehouse is the logical star schema of the stored data. The star schema defines the dimensions of the problem. Often, each dimension of the star schema can be viewed at different abstraction levels. The levels are organized in a hierarchy. Finally, the central entity in the star schema collects the main facts or events of interest. In the case of the BIOBITS project, there are two star schemas.

1. The star built around the *genome composition* facts. It represents the composition of each genome in terms of genes and chromosomes and with reference to the belonging organism.
2. The star schema around *protein* facts. It describes the proteins in terms of PROSITE domains and with respect to the dimensions of phylogenetic classification and metabolic pathways.

The genes and proteins facts are linked by the relationship representing the encoding.

For most of the dimensions, such as genes and phylogenetic classification, the scientific literature already has provided ontologies (e.g., Gene Ontology, GO) and controlled vocabularies (Clusters of Orthologous Groups, COG) that are available in public domain databases and are imported in the system. Another example of available hierarchy on the genes and proteins are the family organizations.

In the following we describe the BIOBITS Data Mart schema (shown in Figure 1) in detail.

Genome Composition. It includes all the relevant information about a genome fragment. Considering a fragment view of the genome, genome composition includes all the known fragments composing a genome: it reports the precise boundaries of the fragments (which depend on the user experience and

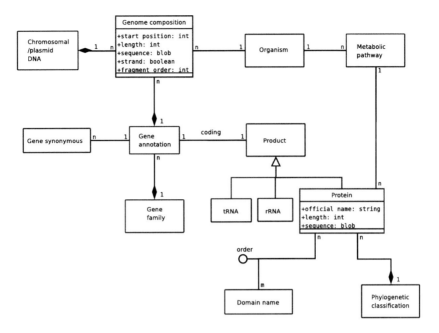

Fig. 1. Star schema of BIOBITS Data Mart

discoveries), the start position and the fragment order with respect to the genome, its nucleotide sequence and strand.

Chromosome/Plasmid DNA. It specifies the localization of the fragment expressed by the number or the name of the corresponding chromosome/plasmid location. Indeed, the genome could be inserted either in a chromosome sequence or in a plasmid sequence.

Organism. It specifies both endosymbiotic and ectosymbiotic bacteria. An organism is identified by the specified identifier, includes the organism scientific name and its classifications in the taxonomy database.

Gene Annotation. It consists in a short report of gene-specific information (identifier and name), comprehensive of a brief description of gene products using both the information reported in Gene Ontology, and the main references stored in `Pubmed`.

Gene Synonymous. It contains all the synonymous names associated to each gene. Genes and proteins are often associated to multiple names; additional names are included as new functional or structural information are discovered. Since authors often alternate between synonyms, computational analysis benefits from collecting synonymous names.

Gene Family. Following the gene classification into families, consistent to the genes biochemical similarity, it reports the family identifiers.

Product. It is a class of the products that genes codify. Products are categorized into in three classes: transfer RNA (tRNA), ribosomal RNA (rRNA) and proteins. Moreover, it reports a pseudogene indication if the gene has lost its coding ability.

tRNA. Transfer RNA is a small RNA molecule that transfers a specific active amino acid to a growing polypeptide chain.

rRNA. Ribosomal RNA is the central component of the ribosome. The ribosome is a complex of ribosomal RNA and ribonucleoproteins.

Metabolic Pathways. It represents pathways which are composed by a set of biochemical reactions. Each pathway represents the knowledge on the molecular interactions and reactions network.

Protein. It refers to protein-specific information (protein identifier and name). A protein is a set of organic compounds (polypeptides) obtained by transcription and translation of a DNA sequence.

Phylogenetic Classification. It consists of Cluster of Ortologous Groups (COG) of protein sequences encoded in a complete genome.

Domain Name. It reports the domains extracted from PROSITE database [22], characterizing the protein sequence. PROSITE consists of documentation entries describing protein domains, families and functional sites.

The relationship among proteins and domains is characterized by the attribute *order* describing how the domains that compose a specific protein are sorted.

3.2 System Architecture

Figure 2 summarizes the main architecture of the BIOBITS system. In the following we focus on objectives and features of the BIOBITS system.

Local and global access to data. The instance of Chado we want to set up will contain both data on genome we shall explicitly produce as part of the project BIOBITS and data retrieved from the biological databases accessible through the Internet. The `Import modules` in Figure 2 will accomplish such a requirements. Concerning the retrieval from Internet, *RRE - Queries* is a GUI wizard, built on the basis of a previously published tool [24], able to query different biological databases like for example GenBank [19] and able to convert the results of the queries into standard formats. Alternatively, we can convert the format of data retrieved from Internet thanks to the scripts available as part of the GMOD project. A remarkable example are those scripts that convert GenBank genes annotations into the Generic Feature Format (GFF), adopted as a standard in the GMOD project. Of course, once data have been retrieved, *Import Modules* update Chado, either on-demand, or automatically, possibly on a regular basis.

An On Line Architecture Mining architecture. One of the advantages of a data warehouse is the ready availability of clean, integrated and consolidated data represented by a multiplicity of dimensions. Once that data are stored in the data warehouse, elementary statistics can be computed on the available facts

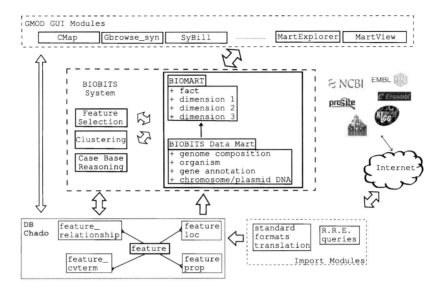

Fig. 2. The architecture of BIOBITS system

and aggregation of measures and frequencies of facts can be immediately computed. The results can be browsed and compared by OLAP primitives and tools. Finally, on these statistics the power of data mining algorithms can be further exploited. This is the On Line Architecture Mining (OLAM) view of a software architecture [20]. OLAM is composed by a suite of data mining algorithms that receive from the client a query for a knowledge discovery task. The request can be answered by the predictive and semi-automatic capabilities of data mining algorithms. In turn, these ones work on the results of an underlying OLAP server that receives the input data from the underlying data warehouse.

For the transformation of the data stored in Chado into the star schema of Figure 1 we exploit BioMart [8], which is a software package available inside GMOD.

Services on Chado and the Star Schema. In Figure 2, associated to both the Chado instance and to the BIOBITS Data Mart we plan to offer two types of services. The first type is implemented on the basis of existing modules of GMOD. Figure 2 highlights them in the uppermost dashed box, named GMOD GUI Modules. The second type of services are internal to the real BIOBITS system: they are shown in Figure 2 inside the central dashed box, named BIOBITS system. Now, we discuss the latter components in detail, putting much emphasis on the features of the software modules that we specifically develop in support to the realization of the goals of the project.

GMOD Graphical User Interface Modules. These modules exploit the available GMOD modules using Chado database to provide the researchers with the tools for comparative genomics needed by the BIOBITS project. GUI modules have also a graphical user interface and allow the user to interact with the system. In particular,

- CMap allows users to explore comparisons of genetic and physical maps. The package also includes tools for maintaining map data;
- GBrowse is a genome viewer, and also permits the manipulation and the display of annotations on genomes;
- GBrowse_syn is a GBrowse-based synteny browser designed to display multiple genomes, with a central reference species compared to two or more additional species;
- Sybil is a system for comparative genomics visualizations;
- MartExplorer and MartView are two user interfaces allowing the user to explore and visualize the stored experimental results and the database content.

BIOBITS system specific modules. The goal of these modules is to allow data analysis under two perspectives that should complement each other and serve for validation.

The first perspective is the one offered by the *Case Base Reasoning* module. It supports efficient retrieval strategies in the context of the search for genomic similarity and syntenies, directly operating on our implementation of the star schema inside BioMart.

The other perspective will exploit tools from Data Mining. We shall use them to perform advanced elaboration on the genomic data. Among the data mining modules we foresee modules for classification, for feature selection and clustering. The latter will be discussed in more detail in this paper, since it has been the first to be integrated into the BIOBITS system. Indeed, one of the main goal of the whole BIOBITS project is to provide the results of fragment alignment tools. Since clustering provides a specifically useful service for the exploration and elaboration of the similarities among genes and proteins, its results could provide to the syntheny tools additional information that would enhance the fragment elaboration.

As a concluding remark, the plan is to develop BIOBITS system specific modules as web-based GUI in order to gain user-friendliness and a good degree of interoperability, similar to current GMOD modules that are able to connect to other modules by standard interfaces.

Of course we shall adhere to the open source philosophy. So, any BIOBITS system specific module will be available as part of the whole project GMOD.

4 Software Modules to Support Researchers' Activities

The main contribution of the BIOBITS project is the development of two GMOD modules to analyse the knowledge stored in the data warehouse. The following section describes the details of these new modules based on Case Based Reasoning and clustering.

4.1 Case-Based Reasoning

Within the BIOBITS architecture, we worked at the design and implementation of an *intelligent retrieval* module, which implements the *retrieval* step of the Case-Based Reasoning (CBR) [2] cycle. CBR is a reasoning paradigm that exploits the knowledge collected on previously experienced situations, known as *cases*. The CBR cycle operates by (1) *retrieving* past cases that are similar to the current one and by (2) *reusing* past successful solutions; (3) if necessary, past solutions are properly *adapted* to the new context in which they have to be used; finally (4) the current case can be *retained* and put into the system knowledge base, called the *case base*. It is worth noting that *purely retrieval* systems, leaving to the user the completion of the reasoning cycle (steps 2 to 4), are very valuable decision support tools [38], especially when automated adaptation strategies can hardly be identified, as in biology and medicine [28]. This is exactly the strategy we are following in the current approach.

Our retrieval module is meant to support comparative genomics studies that represent a key instrument to: (1) discover or validate phylogenetic relationships, (2) give insights on genome evolution, and (3) infer metabolic functions of a particular organism. In the module, cases are genomes, each one taken from a different organism, and properly aligned with the same reference organism. Indeed, the alignment task is a prerequisite in our library. For this reason we start describing the selected sequences alignment strategy, then we detail our module deep down into the cases representation and retrieval.

Sequence Alignment. To deal with the alignment task we rely on BLAST [3]. BLAST is a state-of-the-art local alignment algorithm, specifically designed for bioinformatics applications. It takes as an input a sequence of nucleotides and properly aligns it to a database of strings belonging to (different) organisms of interest.

From a typical BLAST output (Figure 3) one can extract basic information (percentage of the sequence that shows identity and length of the sequence alignment) that can be easily plotted as represented in Figure 4.

Case Representation. From an application viewpoint, it makes sense to convert the *quantitative* similarity values in Figure 4 to a set of *qualitative* levels (e.g. low, medium, high similarity). This provides a "higher level" view of the information, able to abstract from unnecessary details. To perform the conversion, we exploit a semantic-based abstraction process, similar to the Temporal Abstractions (TA) techniques, described in [40,5]. Indeed, in our domain, we consider as the independent variable the symbol position in the aligned strings, instead of the time. As in TA we move from a *point-based* to an *interval-based* representation of data, where the input points are the symbol positions, and the output intervals (*episodes*) aggregate adjacent points sharing a common behavior, persistent over the sequence. In particular, we rely on *state* abstractions [5], to extract episodes associated with qualitative levels of similarity between the two aligned strings, where the mapping between qualitative abstractions

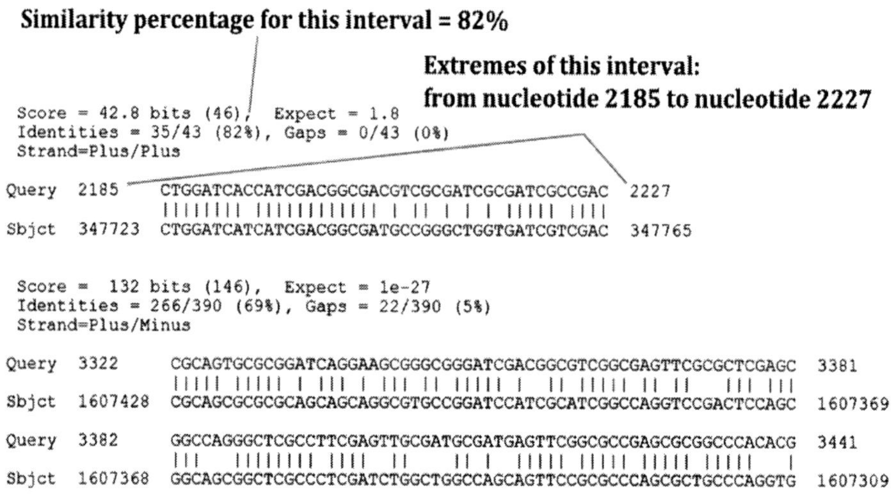

Similarity percentage for this interval = 82%

Extremes of this interval: from nucleotide 2185 to nucleotide 2227

```
Score =  42.8 bits (46), Expect = 1.8
Identities = 35/43 (82%), Gaps = 0/43 (0%)
Strand=Plus/Plus

Query   2185   CTGGATCACCATCGACGGCGACGTCGCGATCGCGATCGCCGAC  2227
               ||||||||  ||||||||||||| | ||  |  |  |||||  ||||
Sbjct   347723 CTGGATCATCATCGACGGCGATGCCGGGCTGGTGATCGTCGAC  347765


Score =  132 bits (146), Expect = 1e-27
Identities = 266/390 (69%), Gaps = 22/390 (5%)
Strand=Plus/Minus

Query   3322    CGCAGTGCGCGGATCAGGAAGCGGGCGGGATCGACGGCGTCGGCGAGTTCGCGCTCGAGC  3381
                |||||  ||||||   |||  |   ||| |  |||||   |  ||||||  ||  |||  |||  |||
Sbjct   1607428 CGCAGCGCGCGCAGCAGCAGGCGTGCCGGATCCATCGCATCGGCCAGGTCCGACTCCAGC  1607369

Query   3382    GGCCAGGGCTCGCCTTCGAGTTGCGATGCGATGAGTTCGGCGCCGAGCGCGGCCCACACG  3441
                |||  ||||||||| ||||   ||   || |   |||||  ||||| |||||  |||||   |
Sbjct   1607368 GGCAGCGGCTCGCCCTCGATCTGGCTGGCCAGCAGTTCCGCGCCCAGCGCTGCCCAGGTG  1607309
```

Fig. 3. BLAST sequence alignment

and quantitative values of the similarity has to be parametrized on the basis of domain semantic knowledge. Semantic knowledge can also support a further refinement of the state abstraction symbols, according to a taxonomy like the one described in Figure 5. Obviously, the taxonomy can be properly modified depending on specific domain needs.

Moreover, our tool allows the representation of the available sequences at any level of detail, according to a taxonomy of granularities, like the one depicted in Figure 6. This granularity change makes sense from a biological point of view: consider e.g. that a region may be conserved among relative organisms, while a specific gene within the region may not. Thus, a high similarity at the region level might be difficultly identified at the level of single genes (as it will be shown in the example discussed in Subsection 5.1).

Notice that the taxonomy of the granularities definition is strongly influenced by domain semantics. For instance, the number of nucleotides which composes a gene depends on the specific organism, and on the specific gene. Domain knowledge also strongly influences the conversion of a string of symbols from a given granularity to a different one, as required for flexible retrieval.

To summarize, *case representation* is obtained as follows. First, an optimal alignment of two nucleotide strings is calculated by BLAST. In particular, for each subsequence of nucleotides, a percentage of similarity with the aligned nucleotide in the paired string is provided. Abstractions on such quantitative levels are then calculated, and allow to convert these values into qualitative ones, expressed as strings of symbols. Abstractions are calculated at the ground level in the symbol taxonomy (and operate also at the ground level in the granularity taxonomy, since they work on nucleotides, see Figure 6). The resulting string of symbols is then stored in the case library as a *case*. Despite the fact that cases are stored as abstractions at the ground level, they could be easily converted at

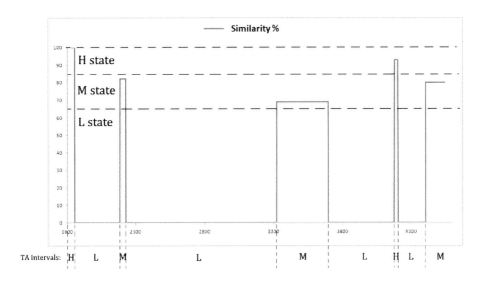

Fig. 4. A graphical visualization of sequence alignment

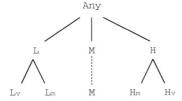

Fig. 5. An example taxonomy of state abstraction symbols; for instance, the high (H) symbol specializes into very high (H_v) and moderately high (H_m)

coarser levels in both dimensions (i.e. the dimension of the taxonomy of symbols, and the one of granularities). Such conversion is the means by which we support flexible case retrieval and will be described below.

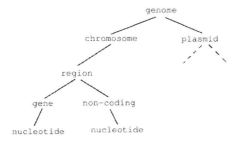

Fig. 6. A taxonomy of sequence granularities

Case Retrieval (query answering). Taking advantage from the *multi-level abstraction* representation introduced above, we support *flexible retrieval*.

In particular, we allow users to express their queries for case retrieval at any level of detail, both in the dimension of data descriptions (i.e. at any level in the taxonomy of symbols) and in the dimension of the granularity.

Obviously, since cases are stored at the ground level in both dimensions, in order to identify the cases that match a specific query, the analyst must provide a function for scaling up (*up* henceforth) two or more symbols expressed at a specific granularity level to a single symbol expressed at a coarser one. Moreover, a proper distance function must be defined.

The data structures described above, as well as the *up* and the distance functions, have to be detailed on the basis of the semantics of the specific application domain. However, we have identified a set of general "consistency" constraints, that any meaningful choice must satisfy, in order to avoid ambiguous or meaningless situations. For instance, we enforce the fact that distance monotonically increases with ordering in the symbol domain.

Moreover, distance "preserves" ordering also in the case in which *is_a* relationships between symbols are involved. For example, the distance between L (low) and M (medium) is smaller than the distance between L (low) and H_v (very high). The exhaustive presentation of such constraints is outside the scope of this paper, but can be found in [29].

In order to increase efficiency, our framework also takes advantage of *multi-dimensional orthogonal index structures*, which allow for early pruning and focusing in query answering. Indexes are built on the basis of the data structures previously described. The root node of each index is a string of symbols, defined at the highest level in the symbol taxonomy, i.e. the children of "Any", as shown in Figure 5, and in the granularity taxonomy. A –possibly incomplete, index stems from each root, describing refinements along the granularity and/or the symbol dimension. An example multi-dimensional index, rooted in the H symbol, is represented in Figure 7. Note that, in the figure, granularity has been chosen as the *leading dimension*, i.e. the root symbol is first specialized in the granularity dimension. From each node of the resulting index, the sequence of the symbols of the node itself is then orthogonally specialized in the *secondary* (i.e. the symbol) *dimension*, while keeping granularity fixed. However, the opposite choice for instantiating the leading and the secondary dimensions would also be possible.

Each node in each index structure is itself an index, and can be defined as a *generalized case*, in the sense that it summarizes (i.e. it indexes) a set of cases. This means that the same case is typically indexed by different nodes in one index (and in the other available indexes). This supports flexible querying, since, depending on the level at which the query is issued, according to the two taxonomies, one of the nodes can be more suited to provide a quick answer.

To answer a query, to enter the more proper index structure, we first progressively generalize the query itself in the secondary dimension (i.e. the symbol taxonomy in the example), while keeping the leading dimension (i.e. granularity

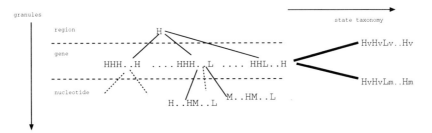

Fig. 7. An example of a multi-dimensional orthogonal index

in the example) fixed. Then, we generalize the query in the other dimension as well. Following the generalization steps backwards, we can enter the index from its root, and descend along it, until we reach the node which fits the leading dimension level of the original query. If an orthogonal index stems from this node, we can descend along it, always following the query generalization steps backwards. We stop when we reach the same detail level in the secondary dimension as in the original query. If the query detail level is not represented in the index, because the index is not complete, we stop at the most detailed possible level. We then return all the cases indexed by the selected node.

It is worth noting that indexes may be incomplete with respect to the taxonomies. Index refinement can be automatically triggered by the storage of new cases in the case base, and by the types of queries which have been issued so far. In particular, if queries have often involved, e.g. a symbol taxonomy level which is not yet represented in the index(es), the corresponding level can be created. A proper frequency threshold for counting the queries has to be set to this end. This policy allows to augment the indexes discriminating power only when it is needed, while keeping the memory occupancy of the index structures as limited as possible.

As a last remark, a number of tools to support comparative genomics studies are already available. For example, the VISTA tool (http://genome.lbl.gov/vista/index.shtml) allows the visualization of pre-computed pairwise and multiple alignments of whole genome assemblies. Our tool, beside alignments visualization, also allows to mine genomes at multiple levels: customized searches can be performed, to retrieve genomes and/or genomic segments matching specific features as described by the query at the desired granularity. Furthermore, thanks to this tool, queries can be performed efficiently and potentially on very large databases. The novelties introduced are exemplified in section 5, with the addition of a performance study.

4.2 Clustering Modules

In this paper we do not go in detail in describing all the predictive and exploratory capabilities offered by data mining algorithms.

The aim of this section is to depict a portrait built on a single example: *clustering*. It offers the possibility to show the benefits in terms of interoperability, extendability and flexibility offered by a modular system built upon a data warehouse in which a multi-dimensional representation of a ground set of facts is stored. On these data, whenever it is needed, a query can be issued by the user in order to retrieve from the data warehouse the values of the interesting subset of dimensions. On this initial set of values multi-level reasoning is possible exploiting the relationships between facts in the knowledge network.

One of the classical aims of clustering is to provide a description of the data by means of an abstraction process. In many applications, the end-user is used to study natural phenomena by the relative proximity relationships existing among the analyzed objects. For instance, he/she compares organisms by means of the relative similarity in terms of the common features with respect to a same referential example. Many Hierarchical Clustering (HC) algorithms have the advantage that are able to produce a dendrogram which stores the history of the merge operations (or split) between clusters. Moreover, the dendrogram produced by a hierarchical clustering algorithm constitutes a useful, immediate and semantic-rich conceptual organization of the object space. As a result HC algorithms produce a hierarchy of clusters and the relative position of clusters in this hierarchy is meaningful because it implicitly tells the user about the relative similarity between the cluster elements. HC approaches help the experts to explore and understand a new problem domain. As regards the exploitation of object distances, clustering algorithms offer immediate and valuable tools to the end-user for the biological analysis.

Co-clustering. A kind of clustering algorithm particularly useful in biological domains is *co-clustering* [14] whose solution provides contemporaneously a clustering of the objects and a clustering of the attributes. Further, often co-clustering algorithms exploit similarity measures on the clusters in the other dimension of the problem: that is, clusters of objects are evaluated by means of the clusters on the features and vice versa. They simultaneously produce a hierarchical organization in two of the problem dimensions: the objects and the features that describe the objects themselves. In many applications both hierarchies are extremely useful and are searched for.

In a more formalized view, a co-clustering algorithm is an unsupervised data mining method that computes a *bi-partition* of a dataset $X \in \mathbb{R}^{n \times m}$. A bi-partition of a dataset is a triple $\langle R, C, \psi \rangle$, where R is a partition of rows (object instances) into $|R|$ subsets, C is a partition of columns (object attributes) into $|C|$ subsets, and ψ is a relation that associate elements of R to elements of C.

An extension of the algorithm based on co-clustering has been obtained by the introduction of *constraints*. Constraints are very effective in many applications, including gene expression analysis [34] and sequence analysis [13], since the user can express which type of biological knowledge leads to the association among the clusters of genes (the *objects*) and the clusters of biological conditions (the *attributes*).

The goal of the constrained co-clustering algorithm is to find a bi-partition such that a given objective function is optimized and a set of user-defined constraints are satisfied. Two kinds of constraints, i.e. *must-link* and *cannot-link*, should be exploited. A *must-link* constraint specifies that two rows (respectively, columns) of X must belong to the same cluster. Conversely, a *cannot-link* constraint specifies that two rows (respectively, columns) of X cannot belong to the same cluster.

In general, the satisfaction of constraints may decrease the theoretical optimum of the objective function. Notice also that the satisfaction of a conjunction of constraints is not always feasible. A constrained co-clustering algorithm works as follows. During each iteration, it associates each row to the nearest row cluster which does not violate any cannot-link constraint. If a row is involved in a must-link constraint the algorithm associates the whole set of rows involved in this constraint to the selected row cluster. Furthermore, it controls that any cannot-link constraint is not violated. This process is iterated until the function reaches a desired value, i.e. its decrease is smaller than a user defined threshold τ. The same process is simultaneously performed over the columns of the matrix.

5 Case Study

The recent efforts of several sequencing projects to explore the genomes of organisms from various lineages have provided great resources for comparative genomics. Since the beginning of the postgenomic era, investigators faced how to manage the rising number of public collections of genomes in novel ways [16]. Other than the public databases where sequences are deposited, more specific data warehouses have been developed [23] were the incorporated data types include annotation of (both protein and non-protein coding) genes, cross references to external resources, and high throughput experimental data (e. g. data from large scale studies of gene expression and polymorphism visualised in their genomic context). Additionally, on such platforms, extensive comparative analysis could be performed, both within defined clades and across the wider taxonomy. Furthermore, sequence alignments and gene trees resulting from the comparative analysis can be accessed. Computational challenges in the field of comparative analyses have been overcome [39]. The developed tools have helped in elucidating the genomic structures of a multiple levels of prokaryotes [6], leading to a much improved understanding of why a bacterial genome is organized in the way it is.

A number of comparative analyses closer to our field of investigation have already shade lights on the characterization of genomes of host-associated and free-living bacteria [41,4,32,11,10]. Novel computational approaches on large scale datasets provide a new viewpoint for whole genome analysis and bacterial characterization. For example, the self-attraction clustering approach allowed classification of Proteobacteria, Bacilli, and other species belonging to Firmicutes [35], whereas the research of protein [18] or genomic [37] signatures have been useful to elucidate the evolutionary relationships among the Gammaproteobacteria and to provide new insights into the evolution of symbiotic diversity, microbial metabolism and host-microbe interactions in sponges.

One major focus of comparative sequence analysis is the search for syntenies. The term synteny is used to mean a set of genes that share the same relative ordering on the genome of different species. In BIOBITS project we are interested on a synteny between several species in order to recognize which are the species more similar to *Ca. G. gigasporarum*. The evolutionary relationships of these genomes may allow the identification of homologous genes and non-coding functional elements, such as regulatory elements and protein domains.

To reach this purpose we exploit the BIOBITS system architecture (shown in Section 3) and the Chado modules described in this paper (see Section 4). To show the reliability of our approach we perform a sequence analysis on a well-known bacterial genus.

5.1 Querying for Synthenies on the Region DCW

Following the Data Mart star schema reported in Figure 1, the data related to a bacterium belonging to the genus of *Burkholderia* (i.e. *Burkholderia xenovorans*) has been extracted. In details, four tables of the Chado database (i.e. *Gene family, Gene annotation, Genome composition,* and *Organism*) are exploited to extract genes belonging to a specific region called Division Cell Wall (DCW). This region is involved in the synthesis of peptidoglycan precursors and cell division. *DCW cluster* is composed of 14 genes: FtsA, FtsI, FtsL, FtsQ, FtsW, FtsZ, mraW, mraY, mraZ, murC, murD, murE, murF, murG. The prominent feature of the *DCW cluster* is that it is conserved with an high (H) similarity in many bacterial genomes over a broad taxonomic range. Specifically, notwithstanding some bacteria belonging *Burkholderia xenovorans* simply miss one of the 14 genes, all of them maintain a high similarity at the DCW region level with their relatives.

Suppose that a user, interested in comparing bacteria on the basis of the DCW cluster content, asks the flexible retrieval system (see section 4) the following query:
$H_v H_v L_v H_v H_v H_v H_v H_v H_v H_v H_v H_v H_v H_v$
looking for the specific bacteria missing the third gene, but very similar to the reference one as regards the other genes. The flexible retrieval system will first generalize the query in the symbol taxonomy dimension (see Figure 5), providing the string: $HHLHHHHHHHHHHH$
and then in the granularity dimension, providing the query H at the region level. Quite naturally, we define the *up* function as:
$up(HHLHHHHHHHHHHH) = H$.

This allows to enter the index in Figure 7 from its root. Then, following the generalization step backwards, a node identical to the query can be found, and the ground cases indexed by it can be retrieved.

Interactive and *progressive* query relaxation (or refinement) are supported as well in our framework. In this situation the distance between the original query and the cases indexed by the other children of the node can be calculated by any distance function which satisfies the constraints illustrated in [29], and quickly described before. Query relaxation or refinement can be repeated several times, until the user is satisfied with the width of the retrieval set. In the *Burkholderia* example, the user may generalize the initial query as an H at the region level, and retrieve also the cases indexed by $HHHHHHHHHHHHH$ at the gene level (the other siblings of $HHLHHHHHHHHHHH$ do not index any real case in this specific situation). The cases indexed by $HHHHHHHHHHHHHH$ can thus be listed, clarifying that their distance from the original query is greater than zero.

Considering the performance of the Case Based Reasoning module, tests have been conducted on databases containing different number of cases. On the left side of Table 1, we report the time elapsed to generate the multi-dimensional indexing structure from the similarity levels generated by BLAST and properly abstracted. The creation times span from 39 seconds to index 2000 cases, to 163 seconds to index 8000 cases. Even if the creation of the structure takes some time, it is necessary to perform this operation only when a new database is installed (or when a significant number of new cases is stored); then the flexible and efficient query mechanism can start running. The right side of Table 1 shows the time elapsed to perform a query, which spans from few milliseconds to query on 2000 cases, to less than one second to query on 8000 cases. These experiments were conducted on an Intel Core 2 Duo T9400 processor running at 2.53 GHz, equipped with 4 Gb of DDR2 RAM.

Table 1. Execution times to build the multi-dimensional orthogonal index (left) and to execute a query (right)

Multidimensional index structure generation from BLAST		Query execution times with multidimensional index	
N. of cases	Structure generation time (s)	N. of cases	Query execution times (s)
2.000	38,969	2.000	0,138
4.000	80,667	4.000	0,333
6.000	121,618	6.000	0,650
8.000	162,241	8.000	0,905

Protein Domains Mining. Beside the investigation of the biological connection at the gene level using the indexing approach, we are able to exploit the *cases* deriving from the case representation to extract new analogies among nucleotide sequences. In details, we query the Chado database to extract all the protein sequences from the obtained *cases*. Then, we use the co-clustering modules to study the domain/motif composition of protein sequences. As it is well known, the modular nature of proteins shows many advantages: it provides an increased stability and new cooperative functions. The usage of protein domains in the determination of the proteins functions has become essential. Several web applications (e.g. Pfam [17], SMART [25], Interpro [31]) are available to provide an overview of the domain architecture of a polypeptide sequence, and the functions that these domains are likely to perform.

Even though the cited tools allow one to submit a set of protein sequences as input, they perform the domain analysis considering each sequence as a single entity. As a consequence, the user can obtain only a *local* view of the domain composition, instead of a *global* view, that may emphasize the domains characterizing the entire proteins set.

This fact suggests the need of an automatic tool that offers the possibility to manage the results in order to highlight the association between domains and proteins. For this purpose the BIOBITS system includes a *de novo* algorithm [13]. It allows the simultaneous association between protein sequences and domains/motifs. In this way we are able to identify a richer set of motifs, each one possibly characterizing only some of the sequences in the whole dataset. The algorithm relies on three steps. First, we generate a prefix tree starting from the sequences in the input dataset. This data structure enables the fast extraction of all the frequent domains of length up to a fixed value w. Then, we exploit a constrained co-clustering algorithm [34] in order to find protein domain classes and the associated protein groups. Finally, we associate the obtained clusters by means of a statistical measure. This measure individuates for each domain cluster the corresponding protein cluster containing it. The statistical measure can associate some protein clusters to any domain cluster, or some protein cluster to more than one domain cluster.

In the presented case study we consider a domain as frequent if it is found at least in the 10% of the sequences given as input, and we set the maximum domain length w equal to 15. The dataset matrix X (defined in Section 4.2) is built using the frequency values stored in the prefix tree. In the definition of the co-clustering constraints, we exploit the Levenshtein distance between two strings. Specifically, we set a must-link constraint on every pair of domains having a distance less than 2. With this limitation, we consider a must-link between two motifs that require only two string operations (i.e. insertion, deletion or substitution) to transform one motif into the other. Otherwise, all the pairs that match by at most two characters are subject to a cannot-link constraint. The stop condition of the co-clustering algorithm is set to be $\tau = 10^{-3}$.

With the above described experimental setting, we performed two types of experiments. In the first experiment we compose the set of input sequences by combing all the *Burkholderia xenovorans*'s protein sequences of genes belonging to *DCW* cluster, stored in table *Protein* of Chado database. The aim of this experiment is the identification of the protein domains common to a *DCW cluster* gene subfamily. We obtain the six motifs reported in logos representation in Figure 8: panel (a) shows the sequence logo representation of the two domains associated to the fts gene family while panel (b) reports the sequence logo representation of the four domains associated to mur gene family.

In order to validate the reliability of our approach we compare our results with respect to the biological knowledge reported in the review by Clyde A. Smith [36]. Smith describes the three domain architectures characterizing the mur ligases. Two of these domains have essentially conserved topology. The author deeply studied the motif composition of one domain, ATPase. It is characterized by a small number of essential structural motifs that include the P-loop motif. The sequence comparisons reported by Smith show the strong conservation of P-loop motif in all four mur ligases. From our analysis we obtain two motifs strictly

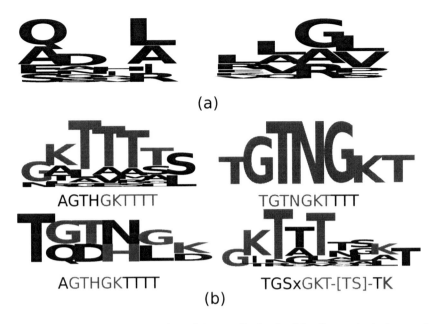

Fig. 8. Sequence logo representation of the motifs obtained by the co-clustering module on *Burkholderia xenovorans*'s *DCW cluster* protein sequences

related to the mur subfamily: in Figure 8(b) we highlight the residues common to the Smith's consensus sequences.

In the second experiment, we exploit the *Phylogenetic classification* table joined to the *Protein* table stored in Chado. The purpose of the second experiment is to extend our analysis to other species of *Burkholderia*. In detail, we single out 13 species:
Burkholderia cepacia, *Burkholderia ambifaria*, *Burkholderia cenocepacia*, *Burkholderia multivorans*, *Burkholderia phytofirmans*, *Burkholderia vietnamiensis*, *Burkholderia glumae* *Burkholderia xenovorans*, *Burkholderia dolosa*, *Burkholderia graminis*, *Burkholderia phymatum*, *Burkholderia rhizoxinica* and *Burkholderia ubonensis*.

The focus of the task is extending the previous association motif/gene subfamily to all *Burkholderia* genus. This kind of analysis is linked to the possibility of understanding if there are one or more domains joined between different species. Figure 9 shows the co-clusters obtained for genes murC, murD, and murE. Our findings confirm that the gene subfamilies are associated with at least one motif and this association is shared with all the orthologous sequences in the *Burkholderia*'s species. The founded domains lead to identify homologous genes, that may catch the evolutionary relationship among a set of genus. The new pieces of information are then stored in the Chado database.

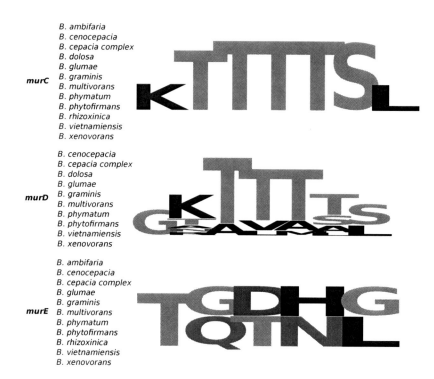

Fig. 9. Co-clusters obtained by performing the co-clustering module on a set of *Burkholderia* species' *DCW cluster* protein sequences

6 Conclusions

In this paper we reported on the on-going **BIOBITS** project whose goal is to extensively develop a computational genomic comparison (known as syntheny) focused on the *Ca. Glomeribacter gigasporarum bacterium* and *arbuscular mycorrhiza* fungi genome.

We presented the software architecture essentially developed over an existing software layer provided by GMOD Community. GMOD system offers powerful data visualization and analysis tools, data warehouse modules, such as **BioMart** and the possibility to exploit import modules for the inclusion of data from the external, public resources. Furthermore, it contains the **Chado** database which presents an extensible and flexible model for any organism species built upon the generic concept of feature which can be customized by the use of types and ontologies.

We presented the logical data representation of the genomic and proteomic components of the biological problem: it has the form of a double star schema - the first one centered around the genetic fragments composing the genome and the second one on the proteins encoded by the genes.

Then, we describe the main software blocks of **BIOBITS system**: a Case-Based Reasoning module and a co-clustering module, which allow the user to retrieve and analyse in a flexible and intelligent way the data coming from the multidimensional star schema. Both these modules complement each other. Case-Based Reasoning and temporal analysis retrieve the information at different abstraction levels, as needed by the analyst. Co-clustering provides a novel information to genetic sequences based on computational data mining algorithms.

In the last part of the paper, we describe a case study showing how these modules inter-operate to provide new information. Interesting results have been obtained, with a confirmation from other research studies. The confirmed reliability of our approach encourages us to continue our research on the endosymbiont bacterium *Candidatus* Glomeribacter gigasporarum.

References

1. Acedb, http://www.acedb.org/
2. Aamodt, A., Plaza, E.: Case-Based Reasoning: foundational issues, methodological variations and systems approaches. AI Communications 7, 39–59 (1994)
3. Altschul, S., Gish, W., Miller, W., Myers, E., Lipman, D.: Basic local alignment search tool. J. Mol. Biol. 215, 403–410 (1990)
4. Bakker, H., Cummings, C., Ferreira, V., Vatta, P., Orsi, R., Degoriciden ja, L., Barker, M., Petrauskene, O., Furtado, M., Wiedmann, M.: Comparative genomics of the bacterial genus Listeria: Genome evolution is characterized by limited gene acquisition and limited gene loss. BMC Genomics 11 (2010)
5. Bellazzi, R., Larizza, C., Riva, A.: Temporal abstractions for interpreting diabetic patients monitoring data. Intelligent Data Analysis 2, 97–122 (1998)
6. Bentley, S., Parkhill, J.: Comparative genomic structure of prokaryotes. Annual Review of Genetics 38, 771–792 (2004)
7. Bianciotto, V., Lumini, E., Bonfante, P., Vandamme, P.: Candidatus Glomeribacter gigasporarum, an endosymbiont of arbuscular mycorrhizal fungi. Int. J. Syst. Evol. Microbiol. 53, 121–124 (2003)
8. BioMart (2003), http://www.biomart.org/
9. Bonfante, P., Anca, I.: Plants, Mycorrhizal Fungi, and Bacteria: A Network of Interactions. Annu. Rev. Microbiol. 63, 363–383 (2009)
10. Carvalho, F., Souza, R., Barcellos, F., Hungria, M., Vasconcelos, A.: Genomic and evolutionary comparisons of diazotrophic and pathogenic bacteria of the order Rhizobiales. BMC Microbiology 10, 1–12 (2010)
11. Commins, J., Toft, C., Fares, M.: Computational Biology Methods and Their Application to the Comparative Genomics of Endocellular Symbiotic Bacteria of Insects. Biomedical Procedures Online 11, 52–78 (2009)
12. Cordero, F., Ghignone, S., Lanfranco, L., Leonardi, G., Meo, R., Montani, S., Roversi, L.: BIOBITS: A Study on Candidatus Glomeribacter Gigasporarum with a Data Warehouse. In: Bohm, C. (ed.) Database Technology for Life Sciences and Medicine Claudia Plant, ch. 10, pp. 203–220 (2011)
13. Cordero, F., Visconti, A., Botta, M.: A new protein motif extraction framework based on constrained co-clustering. In: Proceedings of the 24th Annual ACM Symposium on Applied Computing, pp. 776–781 (2009)
14. Dhillon, I., Mallela, S., Modha, D.: Information-theoretic co-clustering. In: Proceedings ACM SIGKDD 2003, pp. 89–98 (2003)

15. Eilbeck, K., Lewis, S.: Sequence Ontology Annotation Guide. Computational Functional Genomics 5(8), 642–647 (2004)
16. Field, D., Wilson, G., van der Gast, C.: How do we compare hundreds of bacterial genomes? Current Opinion in Microbiology 9, 499–504 (2006)
17. Finn, R., Mistry, J., Schuster-Bckler, B., Griffiths-Jones, S., Hollich, V., Lassmann, T., Moxon, S., Marshall, M., Khanna, A., Durbin, R., Eddy, S., Sonnhammer, E., Bateman, A.: Pfam: clans, web tools and services. Nucleic Acids Res. 34, 247–251 (2006)
18. Gao, B., Mohan, R., Gupta, R.: Phylogenomics and protein signatures elucidating the evolutionary relationships among the Gammaproteobacteria. International Journal of Systematic and Evolutionary Microbiology 59, 234–247 (2009)
19. GenBank (2000), http://www.ncbi.nlm.nih.gov/Genbank/
20. Han, J., Kamber, M.: Data Mining, Concepts and techniques. Academic press, London (2001)
21. Hu, J., et al.: The ARKdb: genome databases for farmed and other animals. Nucleic Acids Res. 29, 106–110 (2001)
22. Hulo, N., Bairoch, A., Bulliard, V., Cerutti, L., Castro, E.D., Langendijk-genevaux, P., Pagni, M., Sigrist, C.: The prosite database. Nucleic Acids Res. 34, 227–230 (2006)
23. Kersey, P.J., Lawson, D., Birney, E., Derwent, P.S., Haimel, M., Herrero, J., Keenan, S., Kerhornou, A., Koscielny, G., Kahari, A., Kinsella, R.J., Kulesha, E., Maheswari, U., Megy, K., Nuhn, M., Proctor, G., Staines, D., Valentin, F., Vilella, A.J., Yates, A.: Ensembl genomes: Extending ensembl across the taxonomic space. Nucleic Acids Research (November 2009), http://dx.doi.org/10.1093/nar/gkp871
24. Lazzarato, F., Franceschinis, G., Botta, M., Cordero, F., Calogero, R.: RRE: a tool for the extraction of non-coding regions surrounding annotated genes from genomic datasets. Bioinformatics 20, 2848–2850 (2004)
25. Letunic, I., Copley, R., Pils, B., Pinkert, S., Schultz, J., Bork, P.: SMART 5: domains in the context of genomes and networks. Nucleic Acids Res. 34, 257–260 (2006)
26. Lumini, E., Ghignone, S., Bianciotto, V., Bonfante, P.: Endobacteria or bacterial endosymbionts? To be or not to be. New Phytol. 170, 205–208 (2006)
27. MAGE Community, MGED Group: MicroArray Gene Expression (MAGE) Project (2000), http://scgap.systemsbiology.net/standards/mage_miame.php
28. Montani, S.: Exploring new roles for case-based reasoning in heterogeneous AI systems for medical decision support. Applied Intelligence 28, 275–285 (2008)
29. Montani, S., Bottrighi, A., Leonardi, G., Portinale, L., Terenziani, P.: Multi-level abstractions and multi-dimensional retrieval of cases with time series features. In: McGinty, L., Wilson, D.C. (eds.) ICCBR 2009. LNCS, vol. 5650, pp. 225–239. Springer, Heidelberg (2009)
30. Moran, N., McCutcheon, A., Nakabachi, P.: Genomics and evolution of heritable bacterial symbionts. Annu. Rev. Genet. 42, 165–190 (2008)
31. Mulder, N., Apweiler, R., Attwood, T., Bairoch, A., Bateman, A., Binns, D., Bork, P., Buillard, V., Cerutti, L., Copley, R., Courcelle, E., Das, U., Daugherty, L., Dibley, M., Finn, R., Fleischmann, W., Gough, J., Haft, D., Hulo, N., Hunter, S., Kahn, D., Kanapin, A., Kejariwal, A., Labarga, A., Langendijk-Genevaux, P., Lonsdale, D., Lopez, R., Letunic, I., Madera, M., Maslen, J., McAnulla, C., McDowall, J., Mistry, J., Mitchell, A., Nikolskaya, A., Orchard, S., Orengo, C., Petryszak, R., Selengut, J., Sigrist, C., Thomas, P., Valentin, F., Wilson, D., Wu, C., Yeats, C.: New developments in the InterPro database. Nucleic Acids Res. 35, 224–228 (2007)

32. Ogier, J., Calteau, A., Forst, S., Goodrich-Blair, H., Roche, D., Rouy, Z., Suen, G., Zumbihl, R., Givaudan, A., Tailliez, P., Medigue, C., Gaudriault, S.: Units of plasticity in bacterial genomes: new insight from the comparative genomics of two bacteria interacting with invertebrates, Photorhabdus and Xenorhabdus. BMC Genomics 11, 1–10 (2010)
33. Osborne, B.: GMOD Community: GMOD (2000), http://gmod.org/wiki/Main_Page
34. Pensa, R., Boulicaut, J.F., Cordero, F., Atzori, M.: Co-clustering Numerical Data under User-defined Constraints. Statistical Analysis and Data Mining (2010)
35. Santoni, D., Romano-Spica, V.: Comparative genomic analysis by microbial COGs self-attraction rate. Journal of Theoretical Biology 258, 513–520 (2009)
36. Smith, C.A.: Structure, Function and Dynamics in the mur Family of Bacterial Cell Wall Ligases. Journal of Molecular Biology 362, 640–655 (2006)
37. Thomas, T., Rusch, D., DeMaere, M., Yung, P., Lewis, M., Halpern, A., Heidelberg, K., Egan, S., Steinberg, P., Kjelleberg, S.: Functional genomic signatures of sponge bacteria reveal unique and shared features of symbiosis. ISME Journal 4, 1557–1567 (2010)
38. Watson, I.: Applying Case-Based Reasoning: techniques for enterprise systems. Morgan Kaufmann, San Francisco (1997)
39. Xu, Y.: Computational Challenges in Deciphering Genomic Structures of Bacteria. Journal of Computer Science and Technology 25, 53–73 (2009)
40. Shahar, Y.: A framework for knowledge-based temporal abstractions. Artificial Intelligence 90, 79–133 (1997)
41. Zucko, J., Dunlap, W., Shick, J., Cullum, J., Cercelet, F., Amin, B., Hammen, L., Lau, T., Williams, J., Hranueli, D., Long, P.: Global genome analysis of the shikimic acid pathway reveals greater gene loss in host-associated than in free-living bacteria. BMC Genomics 11 (2010)

[KD³] A Workflow-Based Application for Exploration of Biomedical Data Sets

Andreas Dander[1,2,3,*,**], Michael Handler[4,**], Michael Netzer[4], Bernhard Pfeifer[4], Michael Seger[4], and Christian Baumgartner[4]

[1] Institute for Bioinformatics and Translational Research, UMIT, Hall in Tirol, Austria
[2] Oncotyrol, Center for Personalized Cancer Medicine, Innsbruck, Austria
[3] Biocenter, Division for Bioinformatics, Innsbruck Medical University, Innsbruck, Austria
andreas.dander@i-med.ac.at
[4] Institute of Electrical, Electronic and Bioengineering, UMIT, Hall in Tirol, Austria
{michael.handler,michael.netzer,bernhard.pfeifer,
michael.seger,christian.baumgartner}@umit.at

Abstract. Based on the biotechnological revolution in the past years, molecular biology has become increasingly data-driven. Knowledge Discovery in Databases, a well-known process in the field of bioinformatics, is supporting the biological research process from data integration, knowledge mining to data interpretation.

This work proposes a new software suite, termed *Knowledge Discovery in Databases Designer* (KD³), covering the complete Knowledge Discovery in Databases process using a workflow-oriented architecture. Three different application-oriented modules are implemented in KD³: First, the *Designer* for designing specific workflows. These workflows can be used by the *Interpreter*, which allows to load and parameterize existing workflows. The *Launcher* encapsulates one dedicated workflow into an independent application to answer one specific biomedical question. KD³ offers a variety of implemented methods, which can be easily extended with new customized components using functional objects. All components can be connected to workflows, which may contain elements of other applications.

Keywords: Knowledge Discovery in Databases, Data Mining, Biomedical Data Exploration, Statistical Analysis, Workflow, Bioinformatics, Software.

1 Introduction

1.1 Motivation

In the past years, huge advances in high-throughput technologies, such as microarrays, mass-spectrometry or high-throughput sequencing have led to a rapid

---
* Corresponding author.
** These authors contributed equally to this work.

growth of amounts of data. Furthermore, publicly available databases are collecting data generated in biological or clinical studies and partly provide free access to this data pool. Examples for such databases are ArrayExpress [27], Gene Expression Omnibus (GEO) [6], or the Sequence Read Archive (SRA) [22].

Knowledge Discovery in Databases (KDD) is a crucial process to support biomedical research and provides a way to manage these enormous data sets, beginning with data integration, subsequent analysis of data, and biological interpretation of findings. One possible application of KDD is in the field of biomarker discovery. Disease biomarkers are indicators which facilitate diagnosis, aid in determining the severity of disease, or allow clinicians to assess the response to therapy.

As standard spreadsheet applications are not able to deal with the complexity of this process and are not suited to handle the massive amount of data, new tools need to be developed. A new workflow-based knowledge discovery tool, termed *Knowledge Discovery in Databases Designer* (KD^3) is proposed here. KD^3 supports the KDD process including data integration, data preprocessing and visualization, and can readily be extended by bioinformaticians and software developers. Three different application-oriented modules of KD^3 are introduced: (i) the *Designer* that allows the development and implementation of new algorithms, and modeling of new sophisticated data analysis workflows, (ii) the *Interpreter* to address a variety of research questions employing pre-defined workflows with the possibility to change parameters, and (iii), the *Launcher*, an automatically generated, tailored stand-alone application, which is used for executing a dedicated workflow to answer one specific question. The purpose of the *Launcher* is to easily distribute designed workflow solutions to research partners.

1.2 Knowledge Discovery in Databases

Fayyad et al. [11] defined the interactive and iterative KDD process as follows: "KDD is the nontrivial process of identifying valid, novel, potentially useful, and understandable patterns in data". This process involves numerous steps with many decisions to be made by the user. In the following these steps are briefly summarized.

1. **Research Question:** Researchers need to develop a basic understanding of the application domain and identify the aim of the KDD process regarding their research question.
2. **Selection of Data of Interest:** In this step researchers need to select a target data set from a study, which contains the data of interest for the discovery process. The subsequent analysis is carried out by focusing on a subset of variables or data samples.
3. **Preprocessing and Cleaning:** Operations such as noise reduction, handling of missing values and outliers, data transformation, homogenization of different domains, or the removal of redundant attributes are performed in this step.

4. **Data Mining:** A key task of the knowledge discovery process is data mining. In this step researchers need to choose between different data mining tasks, such as feature selection, classification (supervised mining), cluster analysis (unsupervised mining), regression analysis or the use of meta models for identifying or classifying novel patterns in the data.
5. **Interpretation/Evaluation:** In the final step researchers need to evaluate and interpret mined patterns or possibly reiterate previous steps if results are not plausible. Subsequently, the new knowledge is documented, reported, or published to interested parties, including approval with established knowledge.

KD^3 [30] has been developed to provide researchers with a tool that covers the entire KDD process including data selection, preprocessing, transformation, data mining, and visualization of findings. Alternative tools that support the entire KDD process, or at least parts of it, comprise different packages in R [13], the open source tools *Orange* [9], *Weka* [15], *KNIME* [8], *RapidMiner* [24], and *Tanagra* [32] as well as commercial products such as *PASW Modeler* [1] from SPSS and *Spotfire Miner* [3] from TIBCO.

In contrast to all previously mentioned tools, KD^3 uses three modules (*Designer*, *Interpreter* and *Launcher*) for different user groups facilitating cooperative work and collaborations. Once a workflow has been designed and validated, a standalone application for the defined workflow, called *Launcher*, can be generated using the *Designer*. After the *Launcher* is generated, it can be distributed to project partners as a single executable Java Archive (jar) file.

2 Implementation

As platform independency was an essential prerequisite for developing the KD^3 application, it has been developed using the programming language Java 1.6, where SWING was selected for designing the components of the graphical user interface (GUI).

KD^3 allows software engineers to simply extend the functionality of the framework by encapsulating new algorithms in *functional objects* (FOs) (e.g. the Load_CSV FO for accessing data of a CSV file shown in both workflows of Fig. 2). Newly developed FOs just need to be copied into a specific directory and KD^3 integrates them like plug-ins by using the Java Reflection API. Additionally, the GUI for each integrated FO is generated automatically, and can be configured with annotations used in the integrated FO.

Due to the modular design of KD^3 other software tools such as *Weka* or *R* can be easily encapsulated within FOs, and incorporated into workflows. Based on this feature, different software products are connected with newly developed algorithms, ingenious established methods, or other applications within a workflow.

2.1 Workflow Execution

A workflow is defined as a network of connected FOs and in order to be valid and executable it has to fulfill the following two criteria: First the workflow must include start points, which are FOs without any preliminary dependency, and second the workflow must not contain cycles, i.e. when traversing the connected FOs, no element must be passed more than once (see Fig. 1).

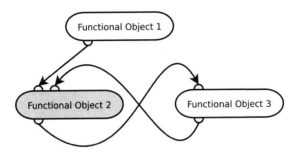

Fig. 1. Cycle in KD^3 workflow. The execution of this workflow is not possible due to a cycle between *Functional Object 2* and *Functional Object 3*.

If a workflow does not fulfill these criteria its execution is not initiated. Otherwise, the launch of a valid workflow triggers all start points. As each FO runs in its own thread, workflow paths can be executed in parallel and are independent from each other.

2.2 Automatically Generated Tailored Stand-Alone Applications

For each workflow a new, independent stand-alone application, called *Launcher*, can be built using the *Designer*. As a result a tailored Java archive file (jar) for executing a specific workflow is available. This jar-file contains all necessary Java classes and libraries for the specified workflow, and is executable on any operating system with installed Sun Java 1.6 or newer. With these customized applications users are able to provide the community with data analysis workflows for specific biomedical questions.

2.3 Implemented Functional Objects in KD^3

This section provides an overview of key FOs used in the mentioned steps of the KDD process. As the selection step is accessing data from different sources, a number of FOs have been developed to assist this step. Various file formats, such as CSV (Comma-Separated Values), XLS (Microsoft® Excel File), or ARFF (Attribute-Relation File Format) can be opened, read and saved. A *Querybuilder* FO is available for connecting different types of relational databases, such as MySQL or PostgreSQL.

A library for basic data table operations has been developed. It supports sorting and merging data tables, and includes FOs for basic transformation processes like transposing or splitting tables, and removing columns and lines of a table.

For the preprocessing step various FOs have been developed to support data transformation by methods such as the z-transformation based on mean and standard deviation, logarithm, and other mathematical scaling methods. In addition, KD^3 provides several FOs for descriptive statistics, outlier detection, hypothesis testing, and normalization.

A suite of FOs is available for the data mining step. For feature selection, algorithms such as the Information Gain [31], ReliefF [20], Biomarker Identifier [7], Stacked Feature Ranking (SFR) [26], have been implemented. In addition the KD^3 library provides FOs for statistical analyses, such as ANOVA [4], Kruskal-Wallis-Test [21], Student's t-Test [4], and many more. Furthermore, there are FOs that support clustering analysis such as KMeans [23], hierarchical clustering [18], OPTICS [5], and others. Moreover, correlation analysis using the JSC library [2] is supported as well.

KD^3 supports users in the evaluation step using different charts for subsequent visualization of these data mining methods, which enables users to find and interpret interesting patterns. Therefore, this step is assisted by diverse visualization techniques [14] such as scatter plots [4], histograms [19], box plots [33], or ROC plots [4].

3 Results and Discussion

A new workflow-based knowledge discovery tool (KD^3) has been developed. KD^3 supports users from different research fields, including data warehouse developers, statisticians, data miners, software engineers, biologists, and physicians. The software package consists of three modules for different application areas: Firstly, the *Designer* which allows an easy development, implementation, and testing of new algorithms, workflows and procedures. Secondly, the *Interpreter* to address specific questions by loading and executing defined workflows. The third module, called *Launcher*, is used for executing a dedicated workflow on a specific problem, and can be generated automatically using the *Designer*. Therefore, the *Launcher* incorporates all necessary resources in one executable jar-File. This application can be simply shared with the research community, and can also be added as supplementary material to a publication.

New algorithms can be integrated into the system by encapsulating algorithms within Java classes. Using defined annotations, the GUI can be generated automatically for each FO, and due to that reason, there is no need to develop a special GUI for each algorithm. Nevertheless, developers with experience in development of user interfaces using the Java API can create own customized GUIs for FOs.

FOs are provided for a variety of tasks for example, different external types of data (flat files and databases) can be loaded and connected. The user is able to select from a collection of preprocessing and transformation procedures, data mining algorithms, and various visualization methods. An example workflow and an introduction on how to design new workflows are available under http://kd3.umit.at/. Additionally, a guideline for developing new FOs is provided at the mentioned webpage ("Hitchhiker's Guide for Generating FunctionalObjects").

3.1 Application Example

This section summarizes a recently published study using KD^3. The aim of this study, published by Netzer et al. [26], was to find breath gas biomarkers that can be used to distinguish between patients with liver disease and healthy controls. Data was generated by analyzing breath gas using ion molecule reaction mass spectrometry (IMR-MS) [17].

Fig. 2 depicts a screenshot of a typical KD^3 application, where both shown workflows were used for identifying breath gas marker candidates in liver disease. The left panel shows a list of available algorithms, either KD^3 specific or integrated from external applications. One visualization window and two different exemplary data analysis workflows are shown in the middle of the screen. The selected FO can be parameterized in the right panel, and after executing the workflow a small preview at the bottom of the right panel displays the result of the selected FO.

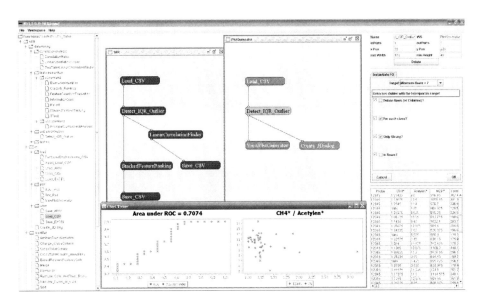

Fig. 2. Screenshot of a typical workflow in KD^3. A tree of available FOs is displayed in the left panel. Information about the currently selected FO (`Detect_IQR_Outlier`) is shown in the right panel (top: location, middle: configuration, bottom: result visualization). Two workflows and two plots as a result of the FO `ViewPlotGenerator` are depicted in the center.

The workflow shown in the left panel, reads the data from a CSV-file and removes outliers in a next step. In the selected FO `Detect_IQR_Outlier` outliers are defined as observations outside the range $[Q_1 - k \cdot IQR; Q_3 + k \cdot IQR]$, where Q_1 and Q_3 are the first and third quartiles, and $IQR = Q_3 - Q_1$ is the interquartile range. The parameter k configures the threshold for defining outliers. This parameter was set to 3 to remove by definition "strong" outliers [26]. The next step in the search for clinically relevant biomarker candidates was to rank 114 features (volatile organic compounds in the breath gas) using the Stacked Feature Ranking (SFR) algorithm. It revealed that CH_4 is the top ranked marker candidate for distinguishing between non-alcoholic fatty liver disease (NAFLD) and controls.

In parallel to SFR the linear correlations between gas compounds were calculated. Note that high correlations between breath gas compounds may indicate uncovered and unexpected relations in chemical reactions useful for interpreting findings. Finally, the two different results for SFR and linear correlation are exported to independent CSV-files. Note that the execution of the independent paths using the `StackedFeatureRanking` and the `LinearCorrelationFinder` is processed simultaneously.

The workflow in the right panel has been used for the visualization of the ROC curve using CH_4 as feature of interest in the left plot and a scatter plot together with Acetylen in the right plot.

Interestingly, with the given workflows new promising breath gas candidates for liver disease were identified, some of which could be verified by literature. In this study, SFR was also compared with other popular data mining algorithms for feature selection (i.e. Information Gain, ReliefF, Biomarker Identifier, and statistical testing), all of them available in KD^3. The results demonstrated that SFR outperforms other methods significantly in terms of the area under the ROC curve (AUC) and is proposed as a powerful tool for the search of biomarker candidates in high-throughput data of mass-spectrometry.

3.2 Discussion

Due to the different steps and possible ways that are necessary for a proper evaluation of biomedical data, workflows for performing analytical and/or computational tasks are important. KD^3 has already been successfully used in several research projects dealing with biomarker discovery [7,16,25,26,28]. One of the main advantages is that the various workflow-design modules enable great usability by providing a preview of results for each step of the workflow. All integrated algorithms can be connected and combined within constructed workflows. Additionally, the application supports parallel calculation of workflow paths to allow optimal performance of each KDD workflow in question.

A further strength of KD^3 is the possibility to develop and implement new algorithms in an easy manner by using so-called *functional objects* (FOs). Newly developed algorithms can easily be compared with already established ones, by creating branches in the workflow.

External software tools and packages, as described in Sect. 1.2, can be simply integrated in KD^3 like plug-ins using FOs. Therefore, different algorithms implemented in different tools can be compared with each other and with newly developed methods. Several data mining methods that already exist in Weka have been integrated using this methodology. Furthermore, advantages of different tools can be combined into one single workflow.

Since generic types are used in the application for transferring data between different FOs, KD^3 can also be applied in other research areas, such as image processing [10], signal processing, chemoinformatics [12], simulation [29], and diverse fields of bioinformatics. At present, more than 100 different FOs supporting the KDD process (i.e., data mining algorithms, graphical representation and data handling tools, tools for executing external processes, query builder components, and others) are available, showing that the development and the integration of new FOs is an ever-ongoing process.

4 Conclusions

The *Knowledge Discovery in Databases Designer* (KD^3) covers the entire KDD process by being able to analyze different types of biomedical data. The proposed concept of three different modules (*Designer, Interpreter, Launcher*) is one of the key features of KD^3, as it provides great flexibility when using the application. Software engineers can leverage already implemented methods, and easily integrate new algorithms. As it is straightforward to integrate external applications into KD^3, our tool simplifies the validation of new algorithms by comparing them with established ones, including methods of several other applications. As a result, KD^3 offers a variety of different methods and applications and allows developers and researchers to encapsulate other software tools.

All components can be connected to tailored workflows, which enable users to answer specific biomedical questions as demonstrated in the example for biomarker discovery in breath gas samples. KD^3 is proposed as a powerful software tool for developers and users which greatly supports the development of new methods, and the discovery of novel knowledge out of biological and clinical data.

Acknowledgements. The authors would like to thank Stephan Pabinger for his constructive comments and proofreading of the final manuscript.

This work was supported by the COMET Center ONCOTYROL and funded by the Federal Ministry for Transport Innovation and Technology (BMVIT) and the Federal Ministry of Economics and Labour (BMWA), the Tiroler Zukunftsstiftung (TZS) and the Styrian Business Promotion Agency (SFG) [and supported by the University for Health Sciences, Medical Informatics and Technology, the Graz University of Technology and Biomax Informatics AG.], and the Austrian Genome Research Program GEN-AU (Bioinformatics Integration Network, BIN III).

References

1. IBM SPSS Modeling Family, http://www.spss.com/software/modeling/
2. Java Statistical Classes, http://www.jsc.nildram.co.uk/
3. TIBCO Spotfire Miner,
   http://spotfire.tibco.com/products/data-mining-applications.aspx
4. Altman, D.: Practical Statistics for Medical Research. Chapman & Hall/CRC (1991)
5. Ankerst, M., Breunig, M., Kriegel, H.P., Sander, J.: OPTICS: ordering points to identify the clustering structure. SIGMOD Rec. 28(2), 49–60 (1999)
6. Barrett, T., Troup, D., Wilhite, S., Ledoux, P., Rudnev, D., Evangelista, C., Kim, I., Soboleva, A., Tomashevsky, M., Marshall, K., et al.: NCBI GEO: archive for high-throughput functional genomic data. Nucleic Acids Research 37(database issue), D885 (2009)
7. Baumgartner, C., Lewis, G., Netzer, M., Pfeifer, B., Gerszten, R.: A new data mining approach for profiling and categorizing kinetic patterns of metabolic biomarkers after myocardial injury. Bioinformatics 26(14), 1745–1751 (2010)
8. Berthold, M., Cebron, N., Dill, F., Gabriel, T., Kötter, T., Meinl, T., Ohl, P., Sieb, C., Thiel, K., Wiswedel, B.: KNIME: The Konstanz Information Miner. In: Studies in Classification, Data Analysis, and Knowledge Organization (GfKL 2007). Springer, Heidelberg (2007)
9. Demsar, J., Zupan, B., Leban, G.: Orange: From Experimental Machine Learning to Interactive Data Mining. Tech. rep., Faculty of Computer and Information Science, University of Ljubljana (2004)
10. Dougherty, G.: Digital Image Processing for Medical Applications. Cambridge University Press, New York (2009)
11. Fayyad, U., Piatetsky-Shapiro, G., Smyth, P.: From data mining to knowledge discovery in databases. AI Magazine 17, 37–54 (1996)
12. Gasteiger, J., Engl, T.: Chemoinformatics: A Textbook, 1st edn. WILEY-VCH, Chichester (2003)
13. Gentleman, R.: R Programming for Bioinformatics. Chapman & Hall/CRC Computer Science and Data Analysis (2008)
14. Grinstein, G., Ward, M.: Introduction to data visualization. In: Fayyad, U., Grinstein, G., Wierse, A. (eds.) Information Visualization in Data Mining and Knowledge Discovery, vol. 1, pp. 21–45. Morgan Kaufmann Publishers Inc., San Francisco (2002)
15. Hall, M., Frank, E., Holmes, G., Pfahringer, B., Reutemann, P., Witten, I.: The WEKA data mining software: An update. ACM SIGKDD Explorations Newsletter 11(1), 10–18 (2009)
16. Herbig, J., Seger, M., Kohl, I., Mayramhof, G., Titzmann, T., Preinfalk, A., Winkler, K., Dunkl, J., Pfeifer, B., Baumgartner, C., Hansel, A.: Online breath sampling with PTR-MS - A setup for large screening studies. In: Proc. 4th Int. Conf. on Proton Transfer Reaction Mass Spectrometry and Its Applications (2009)
17. Hornuss, C., Praun, S., Villinger, J., Dornauer, A., Moehnle, P., Dolch, M., Weninger, E., Chouker, A., Feil, C., Briegel, J., et al.: Real-time monitoring of propofol in expired air in humans undergoing total intravenous anesthesia. Anesthesiology 106(4), 665 (2007)
18. Johnson, S.: Hierarchical clustering schemes. Psychometrika 32(3), 241–254 (1967)
19. Kaminsky, F., Benneyan, J., Davis, R., Burke, R.: Statistical control charts based on a geometric distribution. Journal of Quality Technology 24(2), 63–69 (1992)

20. Kononenko, I.: Estimating attributes: analysis and extensions of RELIEF. In: Bergadano, F., De Raedt, L. (eds.) ECML 1994. LNCS, vol. 784, pp. 171–182. Springer, Heidelberg (1994)
21. Kruskal, W., Wallis, W.: Use of Ranks in One-Criterion Variance Analysis. Journal of the American Statistical Association 47, 583–621 (1952)
22. Leinonen, R., Sugawara, H., Shumway, M.: The sequence read archive. Nucleic Acids Res. [Epub ahead of print] (2010)
23. MacKay, D.: An Example Inference Task: Clustering. In: MacKay, D. (ed.) Information Theory, Inference, and Learning Algorithms, vol. 1, pp. 284–292. Cambridge University Press, Cambridge (2003)
24. Mierswa, I., Wurst, M., Klinkenberg, R., Scholz, M., Euler, T.: YALE: Rapid Prototyping for Complex Data Mining Tasks. In: Ungar, L., Craven, M., Gunopulos, D., Eliassi-Rad, T. (eds.) KDD 2006: Proceedings of the 12th ACM SIGKDD International Conference on Knowledge Discovery and Data Mining, pp. 935–940. ACM, New York (2006)
25. Millonig, G., Praun, S., Netzer, M., Baumgartner, C., Dornauer, A., Mueller, S., Villinger, J., Vogel, W.: Non-invasive diagnosis of liver diseases by breath analysis using an optimized ion-molecule reaction-mass spectrometry approach: a pilot study. Biomarkers 15(4), 297–306 (2010)
26. Netzer, M., Millonig, G., Osl, M., Pfeifer, B., Praun, S., Villinger, J., Vogel, W., Baumgartner, C.: A new ensemble-based algorithm for identifying breath gas marker candidates in liver disease using ion molecule reaction mass spectrometry. Bioinformatics 25(7), 941–947 (2009)
27. Parkinson, H., Kapushesky, M., Kolesnikov, N., Rustici, G., Shojatalab, M., Abeygunawardena, N., Berube, H., Dylag, M., Emam, I., Farne, A., et al.: ArrayExpress update–from an archive of functional genomics experiments to the atlas of gene expression. Nucleic Acids Research 37(database issue), D868 (2009)
28. Pfeifer, B., Aschaber, J., Baumgartner, C., Modre, R., Dreiseitl, S., Schreier, G., Tilg, B.: A data warehouse for prostate cancer biomarker discovery. In: Cohen-Boulakia, S., Tannen, V. (eds.) DILS 2007. LNCS (LNBI), vol. 4544, p. 9. Springer, Heidelberg (2007)
29. Pfeifer, B., Kugler, K., Tejada, M., Baumgartner, C., Seger, M., Osl, M., Netzer, M., Handler, M., Dander, A., Wurz, M., Graber, A., Tilg, B.: A Cellular Automaton Framework for Infectious Disease Spread Simulation. The Open Medical Informatics Journal 2, 58–69 (2008)
30. Pfeifer, B., Tejada, M., Kugler, K., Osl, M., Netzer, M., Seger, M., Modre-Osprian, R., Schreier, G., Tilg, B.: A Biomedical Knowledge Discovery in Databases Design Tool - Turning Data into Information. In: eHealth (2008)
31. Quinlan, R.: C4.5: programs for machine learning. Morgan Kaufmann Publishers Inc., San Francisco (1993)
32. Rakotomalala, R.: TANAGRA: un logiciel gratuit pour l'enseignement et la recherche. In: Sloot, P.M.A., Hoekstra, A.G., Priol, T., Reinefeld, A., Bubak, M. (eds.) EGC 2005. LNCS, vol. 3470, pp. 697–702. Springer, Heidelberg (2005)
33. Williamson, D., Parker, R., Kendrick, J.: The Box Plot: A Simple Visual Method to Interpret Data. Annals of Internal Medicine 110(11), 916–921 (1989)

A Secured Collaborative Model for Data Integration in Life Sciences

Hasan Jamil

Department of Computer Science
Wayne State University, USA
jamil@cs.wayne.edu

Abstract. Life Sciences research extensively and routinely use external online databases, tools and applications for the implementation of computational pipelines. These applications are among the truly distributed and highly collaborative global systems in existence. Since the resources these applications use are designed to serve individual users, they adopt an all-or-nothing model in which users necessarily have to accept the entire response even though only a fraction of the response is relevant. In computational pipelines involving several databases and complex repeat operations, costs due to unnecessary data transmissions and computations could be significant enough to reduce productivity and make the applications sluggish. Since these resources are autonomous, and do not accept user instructions or queries, users are not able to customize their behavior in order to reduce network latency and wasteful computation or data transmission. Obviously, such a resource utilization and sharing model is wasteful and expensive. In this paper, our goal is to propose a new collaborative data integration and computational pipeline execution model for systems biology research. We show that in our envisioned model, arbitrary sites are able to accept user constraints and limited processing instructions to avoid wasteful computation resulting in improved overall efficiency. We also demonstrate that the proposed collaborative model does not breach site security or infringe upon its autonomy.

1 Introduction

Systems biology computational pipelines most often use complicated workflows or application logic that access many distributed databases, online tools and data analysis software in local machines. The distributed nature of these pipelines pose significant data integration challenges that the community has been trying to address for quite some time. The capabilities and the flexibilities available to a computational pipeline largely depends upon the sophistication of the adopted model. The adopted model also determines the overall cost of data analysis for an application which includes the cost of maintenance in addition to the run time query execution cost.

Traditionally, most systems biology applications warehouse remote data and needed tools on local machines by manually reconciling the schema heterogeneity and establishing schema correspondence through schema mapping. Additionally, format transformation functions are implemented to adjust to the formatting needs of various analysis tools and software residing in local machines. Since the data and tools reside in local machines, the user can exercise total control and is able to design applications that are computationally optimized for performance. From a purely computational standpoint, this warehousing model delivers the best run time efficiency among all possible data integration models.

However, the run time performance comes at a huge cost of maintenance, update latency and flexibility. Since data and tools are maintained locally, the user must assume responsibility of ensuring that the warehoused data and tool suite used are current. This maintenance, known as *view maintenance*, is extremely complicated and expensive when updates are frequent at the source. Needless to say that using obsolete versions may lead to incorrect computation and significant waste of resources since the findings will have to be discarded once the version mismatch is discovered. Furthermore, depending on how the source communicates the updates with its users, discovering the changes may mean replacing the entire warehouse[1]. Since the warehouse often needs to employ sophisticated update discovery and maintenance systems to stay current, usually large databases are updated less frequently as it involves downloading substantial amount of remote data over the network. The warehousing approach also suffers from lack of flexibility in choices and scope. Since all data and tools have to be downloaded in local machines, substantial hardware, software and human resources become necessary. This limits what analysis can be supported by the warehouse because any analysis or study that requires resources not maintained in the warehouse becomes impossible.

An alternative has been to write applications by dedicated communication with the data sources, again manually mediating the schema. While this approach removes the physical downloading of the source contents and offers currency, it still requires manual mediation, coping with changes in the source, and writing source specific glue codes that cannot be reused. The basic assumption here is that the sources are autonomous and offers a "use as you see" and hands off support, i.e., the application writer receives no support in any form from the sources other than access.

There has been a significant effort to alleviate the burden on the application writers for this alternative approach by developing libraries in popular scripting languages such as Perl and PHP for accessing and using popular resources such as GenBank [4], UCSC [23], PDB [5], etc. These sources can change their structures without invalidating these libraries, and we have to necessarily write applications using the sources for which tested scripts are available. Consequently, applications that demand change, access to new resources, are transient or ad hoc, and not ready to commit to significant maintenance overhead (as in the former approach) remain ill served.

---

[1] This is possible when the source does not identify which data item has been updated and how.

A slightly more secure approach is to allow access to the database, called the *hidden web*, through predefined web-forms so that users can access the content in a predefined way. Usually the response to such form submissions are returned in HTML. In this environment, users are not only required to use sophisticated applications to fill out forms, gather responses and extract the results in meaningful ways, but also to transport the extracted data to the database for secondary use in their local machines. The interaction follows a client-server model in which the client application interacts with the remote sites individually such that the sites are unaware of any other interactions.

Figure 1 shows the data integration, site interaction or computational pipeline model of the two approaches discussed above. As shown, hidden web sites serve as a "black-box" service provider without any form of collaboration. User applications are responsible for reconciling schema and data heterogeneity, and applications must accept the entire responses these sites generate for queries submitted through the forms. Hence, filtering and preprocessing data for onward submission to another site in the subsequent stage, is the application's responsibility. Such a model ignores the cost of computation and transmission of irrelevant data, which is often significant.

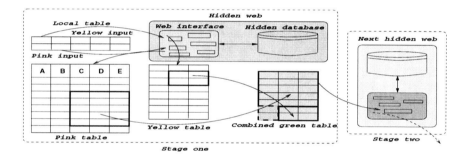

Fig. 1. Schematic view of traditional computational pipelines using hidden web

In this paper, we propose a new collaborative approach to on-the-fly autonomous information integration that removes several of the hurdles in accessing Life Sciences resources, at a throw away cost and without any need for strict coupling or dependence among the sources and applications. Our goal is to preserve site autonomy, and maintain the site's existing security policies, yet offer protocols to reduce unwanted computation and eliminate irrelevant data transmission to improve efficiency. We exploit ideas from query plan generation and distributed transaction processing in databases to develop a collaborative data integration model that is sensitive to site security vulnerabilities. The strategy we adopt is to require sites to accept additional information and instructions relevant to a web interface and provide assistance in carrying out client instructions without compromising established security policies toward improving overall performance of user applications.

2 A Novel On-the-fly Data Integration Model

Traditional data integration approaches based on *global as view* (GAV) or *local as view* (LAV) often become limiting or inapplicable in many emerging applications, especially where conventional assumptions do not apply. For example, to follow either a GAV or LAV based integration model, it is essential that the mapping between the source and global schema is well defined, possibly well in advance. Examples of applications based on LAV and GAV models include BioGuideSRS [7] and MetaQuerier [17] respectively. But many applications, especially in Life Sciences, do not require such stringent view of sources, or even meet the assumptions of GAV and LAV based integration models. Data integration in this domain predominantly takes the form of data heterogeneity as opposed to schema heterogeneity, and a computational pipeline where data moves from one site to the next which possibly includes reconciliation of schema heterogeneity. We explain these issues in the following section using a real life application in Life Sciences in the context of the data integration system LifeDB [6] and its query language BioFlow [21] developed in our laboratory. LifeDB stands apart from other integration systems such as BioGuideSRS and MetaQuerier in the sense that it does not actually advocate a GAV or LAV type schema integration. Instead, LifeDB views data integration from the standpoint of on-the-fly schema mediation, black-box information access through web forms and computational pipeline implementation where no global apriori scheme exists.

2.1 LifeDB Data Integration System: BioFlow by Example

To illustrate the capabilities of LifeDB, we adapt a real life Life Sciences application discussed in [16] which has been used as a use case for many other systems and as such can be considered a benchmark application for data integration. A substantial amount of glue codes were written to implement the application in [16] by manually reconciling the source schema to filter and extract the information of interest. Our goal in this section is to show how simple and efficient it is to develop this application in LifeDB.

The query, or workflow, the user wants to submit is the hypothesis: *"the human p63 transcription factor indirectly regulates certain target mRNAs via direct regulation of miRNAs"*. If positive, the user also wants to know the list of miRNAs that indirectly *regulate* other target mRNAs with high enough confidence score (i.e., $pValue \leq 0.006$ and $targetSites \geq 2$), and so he proceeds as follows. He collects 52 genes along with their chromosomal locations (shown partially in figure 2(a) as the table *genes*) from a wet lab experiment using the host miRNA genes and maps at or near genomic p63 binding sites in the human cervical carcinoma cell line ME180. He also has a set of several thousand direct and indirect protein-coding genes (shown partially in figure 2(d) as the table *proteinCodingGenes*) which are the targets of p63 in ME180 as candidates. The rest of the exploration thus proceeds as follows.

He first collects a set of genes (*geneIDs*) for each of the miRNAs in the table *genes*, from the web site www.microrna.org by submitting one miRNA at a time

(a) genes

miRNA	chromosome
hsa-mir-10a	ch 17
hsa-mir-205	ch 1

(b) sangerRegulation

microRNA	geneName	pValue
hsa-mir-10a	FLJ36874	0.004
hsa-miR-196b	MYO16	0.009

(e) regulation

geneID	miRNA	targetSites	pValue
FLJ36874	hsa-mir-10a	10	0.004
FLJ36874	hsa-mir-10b	3	null
RUNDC2C	hsa-mir-205	8	null
MYO16	hsa-miR-196b	null	0.009

(c) micrornaRegulation

geneID	miRNA	targetSites
FLJ36874	hsa-mir-10a	10
FLJ36874	hsa-mir-10b	3
RUNDC2C	hsa-mir-205	8

(d) proteinCodingGene

Gene	p63Binding
FLJ36874	Y
RUNDC2C	Y
MYO16	N

(f) proteinCodingGeneRegulation

geneID	miRNA	targetSites	pValue	p63Binding
FLJ36874	hsa-mir-10a	10	0.004	Y
FLJ36874	hsa-mir-10b	3	null	Y
RUNDC2C	hsa-mir-205	8	null	Y
MYO16	hsa-miR-196b	null	0.009	N

Fig. 2. User tables and data collected from microRNA.org and microrna.sanger.ac.uk

in the window shown in figure 3(a), that returns for each such gene, a set of gene names that are known to be targets for that miRNA. The site returns the response as shown in figure 3(b), from which the user collects the *targetSites* along with the gene name partially shown as the table *micrornaRegulation* in figure 2(c). To be certain, he also collects the set of gene names for each miRNA in table *genes* from microrna.sanger.ac.uk in a similar fashion, partially shown in table *sangerRegulation* in figure 2(b). This time the column *targetSites* is not available, so he collects the *pValue* values. Note that the scheme for each of the tables are syntactically heterogeneous, but semantically similar (i.e., *miRNA* $\equiv$ *microRNA*, *geneName* $\equiv$ *geneID*, and so on). He does so because the data in the two databases are not identical, and there is a chance that querying only one site may not return all possible responses. Once these two tables are collected, he takes a union of these two sets of gene names (in *micrornaRegulation* and *sangerRegulation*), and finally selects the genes from the intersection of the tables *proteinCodingGene* (that bind to *p63*, i.e., *p63Binding* = 'N') and *micrornaRegulation* $\cup$ *sangerRegulation* as his response.

(a) microRNA.org input form.   (b) microRNA.org returned page.

Fig. 3. Typical user interaction interface at microRNA.org site

To compute his answers in BioFlow using LifeDB, all he will need to do is execute the script in figure 4 to fully implement the application. In a recent work [19], we have shown how a visual interface called VizBuilder for BioFlow can be used to generate this script by an end user without any knowledge of BioFlow in a very short amount of time. It is also interesting to note that in this application, the total number of data manipulation statements used are only seven (statements (2) through (8)). The rest are usual data definition statements. We refer the readers to [6,21] for the meanings of these statements and a discussion on the semantics of this script. But we would like to mention here that ad hoc integration, information aggregation, and workflow design using BioFlow in LifeDB is very simple compared to leading contemporary systems such as Taverna [20] and Kepler [2], and data management systems such as BioMediator [29], BioGuideSRS [7], and MetaQuerier [17].

In the script in figure 4, the statements numbered (1) through (7) are unique to BioFlow and thus of the most importance. The `define function` statements (2) and (4) essentially declare an interface to the web sites at URLs in the respective from clauses, i.e., `microrna.org` and `microrna.sanger.ac.uk`. The `extract` clause specifies what columns are of interest when the results of computation from the sites are available, whereas the `submit` clauses say what inputs need to be submitted. In these statements, it is not necessary that the users supply the exact variable names at the web site, or in the database. The wrapper (`FastWrap`) and the matcher (`OntoMatch`) named in the `using` clause and available in the named ontology `mirnaOntology`, actually establish the needed schema correspondence and the extraction rules needed to identify the results in the response page. Essentially, the `define function` statement acts as an interface between LifeDB and the web sites used in the applications. This statement was first introduced in [11] as the *remote user defined function* for databases where the input to the function is a set of tuples to which the function returns a table. However, the construct in [11] was too rigid and mechanistic, with the user needing to supply all the integration instructions. Actually, it could not use a wrapper or a schema matcher. The user needed to supply the exact scheme and exact data extraction rules. In BioFlow, it is more declarative and intuitive.

To invoke the form functions and compute queries at these sites, we use `call` statements at (3) and (5). The first statement calls `getMiRNA` for every tuple in table `genes`, while the second call only sends one tuple to `getMiRNASanger` to collect the results in tables `micrornaRegulation` and `sangerRegulation`. Statements (6) and (7) are also new in BioFlow, and capture, respectively, the concepts of *vertical* and *horizontal* integration in the literature[2]. The `combine` statement collects objects from multiple tables, potentially with conflicting schemes into one table. To do so, it also uses a key identifier (such as `gordian` [27]) to recognize objects across tables. Such concepts have been investigated in the literature

---

[2] In vertical integration, we collect similar objects from different sources into one set to enlarge the collection (cardinality of a relation) whereas in horizontal integration we extend the information content of objects by acquiring more data from other sources (increase the degree of a relation).

```
process compute_mirna                                                    (1)
{ open database bioflow_mirna;
  drop table if exists genes;
  create datatable genes {
    chromosome varchar(20), start int, end int, miRNA varchar(20) };
  load data local infile '/genes.txt'
    into table genes fields terminated by '\t'
    lines terminated by '\r\n';
  drop table if exists proteinCodingGene;
  create datatable proteinCodingGene {
    Gene varchar(200), p63binding varchar(20) };
  load data local infile '/proteinCodingGene.txt'
    into table proteinCodingGenes fields terminated by '\t'
    lines terminated by '\r\n';
  drop table if exists micrornaRegulation;
  create datatable micrornaRegulation {
    mirna varchar(200), targetsites varchar(200), geneID varchar(300) };
  define function getMiRNA
    extract mirna varchar(100), targetsites varchar(200),
    geneID varchar(300)
    using wrapper mirnaWrapper in ontology mirnaOntology
    from "http://www.microrna.org/microrna/getTargets.do"
    submit( matureName varchar(100), organism varchar(300) );      (2)
  insert into micrornaRegulation
    call getMiRNA select miRNA, '9606' from genes ;                (3)
  drop table if exists sangerRegulation;
  create datatable sangerRegulation {
    microRNA varchar(200), geneName varchar(200), pvalue varchar(200) };
  define function getMiRNASanger
    extract microRNA varchar(200), geneName varchar(200),
    pvalue varchar(30)
    using wrapper mirnaWrapper in ontology mirnaOntology
    from "http://microrna.sanger.ac.uk/cgi-bin/targets/v5/hit_list.pl/"
    submit( mirna_id varchar(300), genome_id varchar(100) );       (4)
  insert into sangerRegulation
    call getMiRNASanger select miRNA, '2964' from genes ;          (5)
  create view regulation as
  combine micrornaRegulation, sangerRegulation
    using matcher OntoMatch identifier gordian;                    (6)
  create view proteinCodingGeneRegulation as
  link regulation, proteinCodingGene
    using matcher OntoMatch identifier gordian;                    (7)
  select *
    from proteinCodingGeneRegulation
    where pValue <= 0.006 and targetSites >= 2 and p63binding='N'; (8)
  close database bioflow_mirna; }
```

Fig. 4. BioFlow script implementing the process

under the titles record linkage [30] or object identification [24]. For the purpose of this example, we adapted GORDIAN [27] as one of the key identifiers in BioFlow. The purpose of using a key identifier is to recognize the fields in the constituent relations that essentially make up the object key[3], so that we can avoid collecting non-unique objects in the result. The `link` statement, on the other hand, extends an object in a way similar to join operation in relational algebra. Here too, the schema matcher and the key identifier play an important role. Finally, the whole script can be stored as a named *process* and reused using BioFlow's `perform` statement. In this example, line (1) shows that this process is named `compute_mirna` and can be stored as such for later use.

2.2 LifeDB Data Integration Model

It may be fairly obvious from the script above that LifeDB adopted a model similar to the one shown in figure 1, in which the application assumes the responsibility of coordinating the query execution plan by individually interacting with the participating sites that respond only to predefined queries by computing one of many possible views. Although we have collected two similar sets of objects from sanger.org and microrna.org, the global schema is not predefined. Finally, we have accessed these two sites to create a unified view of their contents based on their local scheme. In other words, we blurred the distinction between GAV and LAV. Interestingly, the combine statement in (6) actually removes the restrictions imposed by GAV and LAV because it retains all possible information under these two integration models individually, whereas the link statement in (7) extends objects in source databases in a way not possible under GAV and LAV as shown in figure 5.

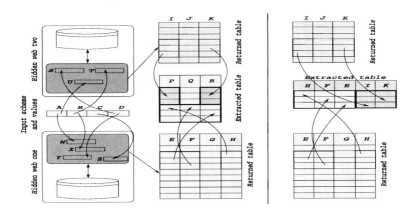

Fig. 5. Vertical (left) and horizontal (right) data integration in LifeDB

---

[3] Note that object key in this case is not necessarily the primary keys of the participating relations.

Unlike GAV and LAV, the model in figure 5 uses the queries as the driver of integration without a required global scheme using a "best effort" approach similar to [8]. In this approach, the input/output behavior of the web forms (the interface to the hidden web) are matched with the query scheme (the scheme of the input relation and the variable list in the select clause) to the extent possible. In the example in figure 5, the input scheme $\{A, B, C, D\}$ is mapped to the form entries of first hidden web form $\{W, X, Y, Z\}$. The site responds with a table with scheme $\{E, F, G, H\}$, which in turn is mapped to the select variables $\{P, Q, R\}$. Once the schema correspondence is established (shown with arrows), and query selection conditions are applied, only a subset of the rows and columns are selected for the combined table (shown with arrows). From the second hidden web, we follow a similar procedure. The only difference is that we are now only able to map a subset of our input values and our select clause variables to the web form. Under the best effort assumption, currently we require that all the input values need to be mapped to input form entries, but not all the select clause variables need to be found in the site response. Consequently, the table extracted from a site response may not have all the columns, and often needs to be padded with null values. In this example, it is shown that the table extracted from site two did not have an identifiable scheme either. The scheme $\{I, J, K\}$ has been assigned first using a best effort annotation algorithm and then mapped to the overall scheme $\{P, Q, R\}$. We are able to map only $\{I, K\}$ to $\{P, R\}$ respectively, and pad Q with nulls.

This approach requires that we "identify" objects, not tuples, and collect the objects with as much information as possible. We have adopted the traditional approach to object recognition [28,15,30] and apply a procedure to extract objects as opposed to tuples in a relation. In this sense, our approach to the combine operation is entirely different from the traditional union or outer union operation [18], but resemble the fusion of outer union with entity resolution. The link operation as shown in the right of figure 5 has a similar approach to schema mapping and table extraction. However, this operation is distinct from a natural join or theta join operation as it uses discovered objects as join conditions, which is potentially different from the joining columns (same name attributes).

2.3 Computational Pipeline Coordination and Cost Model

In most scientific computational pipelines, as well as in LifeDB, the coordination model is relatively rudimentary and straightforward. In these models, the application directs the control flow, communicates with the sites in series, collects the responses, and moves from stage to stage as shown in figure 1, and abstracted in figure 6 in which the application does not expect site cooperation beyond the services they already provide. The application assumes the responsibility for schema mapping, table extraction, missing column name annotation, object recognition, table consolidation and filtration. Therefore, network transmission time needed to send and receive result or intermediate relations can be considered the dominant cost of computation as communication speed is significantly

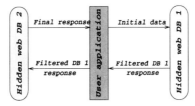

Fig. 6. LifeDB computational pipeline coordination model

lower than local computation[4]. This cost can be measured in the number of bytes in a tuple that are transmitted in either direction.

Previous research has developed sophisticated transaction and coordination models for distributed query processing, where data integration is not a serious consideration. In other words, the distributed scheme is well understood and a tight coupling of the database engines is assumed. In our approach, these assumptions do not hold, and databases are fully autonomous and mostly opaque to the outside world even though they cooperate in a client-server type model. Our goal in this paper is to preserve the opacity of sites, yet make them accept instructions from clients toward reducing computational cost at the site, as well as reduce network communication cost by eliminating unnecessary data transmission without compromising the site security model. Although the sites will now assume some of the functions of the client, or the application, the overall computational efficiency will be improved.

3 A Collaboration Model for Computational Pipelines

To curtail the cost of unnecessary network transmission during a computational pipeline execution, we must receive some form of cooperation from the participating hidden web sites that traditionally do not accept anything but form submissions. Forms typically accept user input data and query conditions. These inputs are then converted into a well structured query – usually an SQL query when the underlying database management system uses a relational engine. To improve application performance, there are two levels of cooperation a site potentially can provide without compromising its security. To retain autonomy and independence, all sites essentially adopt the black-box behavior. In other words, it still remains the responsibility of the application to decide how to use the site resources and what semantics to assign to the responses the site generates. In the next two section we discuss how the hidden web sites could help improve application efficiency by accepting data filtration and process coordination instructions from applications toward reducing data transmission cost, and client side post-processing cost without compromising server side performance.

---

[4] Traditional distributed database models also use this simplified cost model for query cost evaluation.

3.1 Reduced Data Transmission through Collaboration

First, the site can assist by returning a table with no redundant tuples from an application's viewpoint. The basic idea is depicted in figure 7 in which we show that the hidden web site now becomes responsible for the functions of the application in figure 1. In other words, the site now sends one optimized response as opposed to a series of large tables corresponding to each form submission. That means, a site must accept two pieces of information – a mapping from application query scheme to the scheme of the response table the site will generate, and a filter condition that the application would like to enforce.

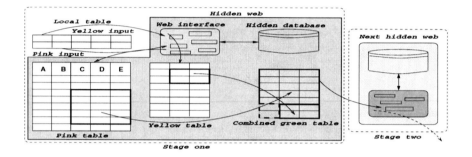

Fig. 7. Modified computational pipeline using cooperative hidden web resources

There are at least two barriers to this collaboration scheme. Ideally, an application does not know the scheme of the returned table before submitting the form as it is following an on-the-fly integration protocol. Therefore, it cannot send the mapping information for the table to the site, and must request the site to carry out the mapping. Since mapping is often an application specific feature, and potentially can determine the application semantics, the choice of the mapping algorithm must reside with the application. Unfortunately, since a site cannot possibly host all possible mapping algorithms nor can it accept an executable for security reasons, the application must either send it, or help the site find it. To avoid a potential security breach, we believe that a trusted site can host possible mapping algorithms that a server site can download from to carry out the mapping, using the application's mapping of choice. The resulting model will have the appearance of the schematic diagram in figure 8, called *synchronous coordination*. In this coordination model, the application acts as the central communicator and query execution manager.

Alternately, the application or client site could host a mapping service of its own to assist the server site to compute the mapping. This approach is interesting when the application is using exotic schema matchers specific to the application and most other application will not benefit from it. Thereby depositing the mapper at any trusted site will result in little or no benefit. In this paper, however, we will follow the first approach and assume that a trusted site exists from which a site can download the schema matching algorithm.

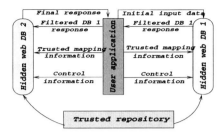

Fig. 8. Synchronous coordination

To reduce the size of the final result, the simplest solution is to apply the selection condition on the view created by the site and project out the unneeded columns after schema mapping. Although it optimally reduces the size of the table to be transmitted, it misses the opportunity to help the site reduce the computational cost. A better approach would be to push down the application constraints to the site query so that unnecessary tuples are not generated in the first place. Such cost reduction will help the sites balance the additional cost they incur to compute application queries. We will revisit this issue in section 4 to discuss secured collaboration.

3.2 Coordination Models for Collaboration

The synchronous coordination model in figure 8 still requires the site to forward the optimized table to the coordinating site, i.e., the application. If the application simply needs to forward the table returned by site 1 to site 2, as shown in figure 8, then we can further improve the performance by requesting site 1 to simply forward or submit the table to site 2 directly. That way we are able to save one complete transmission. Again for this to be successful, we will need a mapping from the table scheme to the site 2 web form scheme, and communicate a possible mapping information to site 2 as we did in the case of site 1. The collaboration model under this scenario will be significantly different than the previous approach. The idea of the new coordination model can be described using the diagram in figure 9 in which we show a distributed coordination.

In the distributed coordination model, called the *asynchronous coordination model*, an application sends not only mapping information, input data and selection condition to the site, it also sends the query plan, and other control information. The site, upon completion of its portion of the computation in the query plan, forwards the required portion of the remaining query plan and control information to the next site alongwith the table it produced as input. Optionally, the site may inform the application the status of the computation to improve coordination. The complexity and sophistication of such an approach is not the subject of this article. Instead we are merely interested in proposing the model, and developing the detailed protocols as a complete research to be reported elsewhere. The basic and abstract model of coordination is shown in

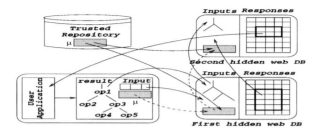

Fig. 9. Asynchronous collaboration model. Solid arrows show actual data transfers and dashed arrows show control information transfers.

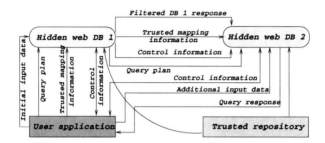

Fig. 10. Asynchronous coordination

figure 10, which differs with the synchronous model in that each site now shares part of the coordination and query plan, and the application no longer has full knowledge of the state of computation.

A hybrid of synchronous and asynchronous models may become essential when the application needs to augment the pipeline with site specific additional information (in the form of data or control information), needs to have a monitoring or supervisory role, or if it is simply expensive to send the whole or part of the information down the chain of controls at each site. The longer or more complex the pipeline, the more control information will be required to be transmitted. A more sophisticated coordination similar to distributed transaction processing and coordination may become necessary to be optimal.

4 Reducing Query Cost Using Parameterized Views

Usually each site responds to a form submitted by a client by computing a view over its database back-end. Even though it uses a structured database, it often exposes the computed view as a semi-structured document, usually in HTML, over the internet so that the content is viewable based on a style sheet. The scheme of the exposed view is often transformed as well for presentation purposes, or to protect the structure of the internal database scheme. Many hidden web sites often remove the scheme of the view because the data is expected to be consumed by a human end user who is able to gather the meaning naturally. The diagram in figure 11 explains this black-box interaction model of the sites where arrows show the scheme mappings. This layered separation of the structure of the database makes it difficult to design an autonomous system to manipulate view definition without site cooperation. In this section, our goal is to devise a mechanism to allow some degree of view manipulation without security breaches.

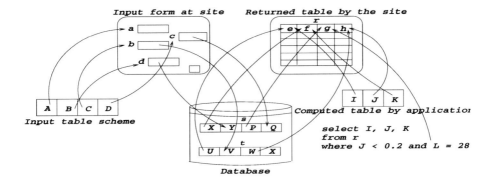

Fig. 11. Hidden web site and application interaction model

Let us assume that the site uses a query Q over the database tables s and t as shown below to compute the view presented in figure 11 that the application calls table r.

select X, P, W
from s, t
where $s.X = t.X$ and $U < 17$ and $Y = d$ and $V = b$ and $Q = c$

To be able to safely push a selection condition $\theta = J < 0.2 \wedge L = 28$ as shown in the SQL query in figure 11, we must rewrite query Q as Q' as follows:

select X, P, W
from s, t
where $s.X = t.X$ and $U < 17$ and $Y = d$ and $V = b$ and $Q = c$ and
 $W < 0.2$ and $P = 28$

Rewriting Q into Q' is relatively straightforward when the mapping among the variables are known, and the query condition θ is given. In fact, the server interfaces routinely rewrite queries based on form input entries and executes the modified queries. In our case, the base query Q_b is perhaps of the form (using the syntax and convention in Roichman and Gudes [26])

select X, P, W
from s, t
where $s.X = t.X$ and $U < 17$ and $Y = :d$ and $V = :b$ and $Q = :c$

which is rewritten into Q by replacing $:x$ variables with the corresponding values x from the interface slots. Roichman and Gudes [26] observed that uncontrolled use of such interface parameters to rewrite queries could lead to a security breach called *SQL injection*. To illustrate this security vulnerability, consider the SQL query in the form of a string

strSQL = "select *Salary*
from *Salary_Table*
where *Employee_No* = *123* and *Salary_Date* = '" + *dateParam* + "'"

where *dateParam* is a user supplied string to be concatenated to form the final SQL query. Now if the user will enter *01/2011'* or *'1'* = *'1*, then we'll get:

select *Salary*
from *Salary_Table*
where *Employee_No* = *123* and *Salary_Date* = *'01/2011'* or *'1'* = *'1'*

This expression will now return salary information of all employees because the disjunct *'1' = '1'* is a tautology. Therefore, there is significant risk in allowing even apparently harmless rewriting of the view definition. Roichman and Gudes suggested a security protocol using passkey to deal with this breach. Unfortunately, a passkey is not an acceptable solution in scientific databases where a user could land, without notice, on a site virtually from anywhere in the globe, and maintaining any form of passkey or adopting a grant/revoke type of authorization is not feasible. Therefore, we must devise a non-intrusive yet effective method for accepting user constraints without any security breach.

4.1 Passing Parameters to Views

Deductive languages such as Datalog [10] are arguably richer and more powerful in the way views can be manipulated as part of queries. Traditionally, SQL views do not accept parameters, and hence any filtration must be carried as a remedial operation. This is one of the reasons why in practical web interfaces and other user directed query evaluation engines, a literal query composition based approach is adopted to accept user constraints in remote databases at significant risk for security breaches. In the next section, we review the limitations of relational views as opposed to views in deductive languages on intuitive grounds.

Inadequacy of Relational Views. SQL views are traditionally used to create a restricted subset (called relational restrictions) of a set of database relations that possibly involve complex operations such as aggregation and other specialized functions. Views are useful for numerous important data processing activities. For example, views can be used to create the user's universe of discourse, enforce controlled access or authorization policies, implement predefined queries, pose complex queries in a stepwise fashion, and implement stored procedures. However, views are essentially static and cannot be used for changing needs. That is, views are defined with specific applications in mind and hence, are not customizable. More specifically, the conditions in the where clause are fixed and cannot be modified without changing the program. This rigidity can be considered a severe limitation of SQL in comparison to other procedural languages such as C and PL/SQL, and declarative languages such as Datalog. In the remainder of this section, we discuss the insufficiency of SQL views as an advanced database programming construct in an attempt to justify the development of the concept of *parameterized views* and *relational unification*.

Consider a simple student database schema as shown in figure 12, and consider two derived or virtual attributes *income* defined as shown, using dotted lines in the entity sets employee and student, and *bonus* defined only in employee. The idea here is that the *Income* attribute in employee is more specialized and hence overrides the definition in student, in a way similar to object-oriented systems. However, since grad and employee are specializations of student, they inherit all the attributes – static (shown using solid lines) and virtual, of student. Assume also that the SQL DDL statements in figure 13 are used to define the schema of these relations.

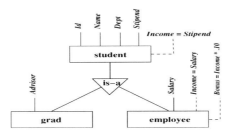

Fig. 12. Student database ER schema

There could be several implementation strategies to map this ER-like schema into a relational schema. One possible representation is shown in figure 14 where tuples are placed at the level where they actually belong, instead of splitting them in general and specialized entity sets. Notice that the *income* and *bonus* attributes are not shown as they are virtual and could be implemented using SQL query or view definitions. Also, *Joe* and *Marni* are both graduate students and employed as teaching assistants, and so they appear in both grad and employee tables with appropriate attributes.

```
create table student      create table employee     create table grad
( Id      integer,        ( Id      integer,        ( Id      integer,
  Name   char(10),          Name  char(10),           Name    char(10),
  Dept    char(10),          Dept   char(10),          Dept     char(10),
  Stipend integer )          Salary integer )          Stipend integer,
                                                       Advisor char(10) )
```

Fig. 13. Relational schema definition

Now let us consider a view v_1 called ta_info in figure 15 constructed from the relations grad and employee to respond to queries that include *income* and *bonus* as attributes in addition to the other static attributes of grad and employee. We implement the views v_2: ta_income and v_3: ta_bonus as shown in figure 15 corresponding to *income* and *bonus* and use them to define the view ta_info. What is worth noting here is that the ta_info view is concretely defined for *CS* teaching assistants only. It is not possible to find out information about *all* or for teaching assistants from *PH* department using ta_info view because this view has already been filtered away. We could redefine ta_income view shown as v'_2 in figure 15 and then execute a query over ta_info as follows:

> select *
> from ta_info

or

> select *
> from ta_info
> where Dept = "PH"

The view ta_info now computes information for all departments and we are able to choose. Of note here is that in the latter approach, we did not have the opportunity to push the selection constraint $Dept = "PH"$ to the view definition for ta_income, and thus more time was spent computing redundant information that is discarded eventually.

student			
Id	Name	Dept	Stipend
27	Oishii	CS	1000
45	Sharaar	EE	800

grad				
Id	Name	Dept	Stipend	Advisor
39	Susan	CS	1200	Carter
88	Tony	ME	2800	Miller
12	Joe	PH	1200	Pots
82	Marni	CS	1000	Renee

employee			
Id	Name	Dept	Salary
61	Brenda	CS	3600
22	Smith	PR	2800
12	Joe	PH	1600
82	Marni	CS	1800

Fig. 14. Student database instance

On the contrary, systems such as Datalog offer a neater way of managing such scenarios through opportunities to optimize naturally. For example, consider the set of Datalog rules below.

> r_1 : *ta_info(Id, Name, Dept, Stipend, Advisor, Salary, Income, Bonus)* ← *grad(Id, Name, Dept, Stipend, Advisor)*, *ta_income(Id, Income, Dept)*, *employee(Id, Name, Dept, Salary)*, *ta_bonus(Id, Bonus)*.
> r_2 : *ta_income(Id, Income, Dept)* ← *grad(Id, _, Dept, Stipend, _)*, *employee(Id, _, Dept, Salary)*, *Income = Salary + Stipend*.
> r_3 : *ta_bonus(Id, Bonus)* ← *ta_income(Id, Income, _)*, *Bonus = Income * .10*.

Recall that these intensional database (IDB) rules in Datalog are evaluated on demand following a query, whereas the views in SQL are computed right away

v_1:
create view ta_info as
select Id, Name, Dept, Stipend, Advisor, Salary, Course, Income, Bonus
from $employee$ as e, $grad$ as g, ta_income as i, ta_bonus as b
where e.Id = i.Id and i.Id = b.Id and b.Id = g.Id

v_2:
create view ta_income as
select Id, Income = Stipend + Salary, Dept
from $employee$ as e, $grad$ as g,
where e.Id = g.Id and Dept = "CS"

v_3:
create view ta_bonus as
select Id, Bonus = Income * .10
from ta_income

v_2':
create view ta_income as
select Id, Income = Stipend + Salary, Dept
from $employee$ as e, $grad$ as g,
where e.Id = g.Id

Fig. 15. View definitions on student database

and maintained. In SQL, only the queries get executed at run time in an ad hoc fashion. In this way, Datalog rules (although equivalent to views in SQL) behave like procedures, and the queries take the role of procedure calls that provide appropriate variable adornments. Given the rules r_1 through r_3, we now have many choices to ask queries that were not possible with SQL views so easily.

Now consider the queries below and their SQL and Datalog representations as shown in figure 16 based on view definitions v_1, v_2' and v_3, Datalog rules r_1 through r_3, and their responses in figure 17.

q_1 : List names of all CS teaching assistants and their bonuses.
q_2 : List names of all CS teaching assistants, their salaries and bonuses.
q_3 : List names of all physics teaching assistants and their bonuses.

What is interesting to note is that to evaluate the Datalog version of the query q_1 through q_3, the constants "CS" and "PH" propagate throughout the rules due to *sideways information passing* and *unification* [10,3] and help optimize the query evaluation. A similar optimization is not possible in existing SQL because variable adornments cannot be passed from one view to the next. As a consequence, in SQL, we are forced to write a separate view for every possible optimized invocation or to write a general enough view and select a desired subset as needed thus rendering the process restrictive, inefficient, and expensive.

The queries in figure 16 also expose several interesting properties of SQL and Datalog that are relevant for our ensuing discussion on parameterized views. Query q_1 shows that SQL supports arity and position independence whereas in Datalog, name independence is supported while they both produce the relation in figure 17 (a) as an answer. By arity and position independence we mean that

```
q₁: SQL      : select Name, Bonus
                from ta_info
                where Dept = "CS"
     Datalog : ? ta_info(_, X, "CS", _, _, _, _, Y).

q₂: SQL      : select Salary, Name, Bonus
                from ta_info
                where Dept = "CS"
     Datalog : ? ta_info(_, X, "CS", _, _, S, _, Y).

q₃: SQL      : select Bonus, Name
                from ta_info
                where Dept = "PH"
     Datalog : ? ta_info(_, X, "PH", _, _, _, _, Y).
```

Fig. 16. SQL and Datalog versions of queries q_1 through q_4

Name	Bonus
Marni	280

(a)

Name	Salary	Bonus
Marni	1800	280

(b)

Name	Bonus
Joe	280

(c)

Fig. 17. (a) and (b) - SQL responses for q_1 and q_2, and (c) Datalog response for q_3

for a given scheme $r(R)$ where $R = \{A, B, C\}$, select A, B from r where A=a is allowed even though the degree of R is 3, and will produce identical response to the query select B, A from r where A=a. However, select X, B from r is not allowed and will result in a syntax error because there is no attribute named X. That means, the attribute name used in the query must appear in the scheme, but its position listed is immaterial. In contrast, in Datalog, ? r(A, B) is not allowed (technically speaking, it will produce nothing – a logical `false`), but ? r(A, B, C) or ? r(A, B, _) are. That means, the degree or arity of the predicate and the relation must match. However the variable names (or attribute name) does not matter at all since the queries ? r(A, B, C), ? r(B, A, C) and ? r(X, Y, Z) will produce identical responses including the order of the columns. In contrast, although select A, B, C from r and select B, A, C from r will produce mathematically identical relations, the order of the attributes will be different.

4.2 Necessity of Parameterized Views

It is our contention that if we could mimic Datalog's side ways information passing [3] to be able to propagate attribute values from one view to another as parameters, we could actually make remote hidden web sites accept user constraints in the form of constants. If the supplied constants can be used to filter unwanted data, we could improve the computational efficiency at the server side.

In other words, we could impose further restrictions on the view implementing a hidden web query form by pushing selection conditions inside. The goal here is to facilitate generic view definitions that can be evaluated on demand in a way similar to Datalog using variable bindings, as shown in example 1.

Example 1. Consider simulating the Datalog version of the views ta_info and ta_income. If we could write these views as shown in figure 18, and supply the constant "CS" when we reference ta_info, we could achieve our goal. Notice the use of a call statement in the from clause of view v_4: ta_info that passes the *Dept* value as D to view v_5: ta_income which utilizes it in its where clause for selecting the appropriate subset of tuples. The ta_info view is *activated* or *fired* through a similar call statement query q_4 as shown in figure 18. □

v_4:
    create view *ta_info* with
    parameter D as
    select *Id, Name, Dept, Stipend, Advisor, Salary, Income, Bonus*
    from *employee* as e, *grad* as g, call *ta_income* with (D'/D) as i,
    ta_bonus as b
    where e.Id = i.Id and i.Id = b.Id and b.Id = g.Id and Dept = D

v_5:
    create view *ta_income* with
    parameter D' as
    select *Id, Income* = *Stipend* + *Salary*
    from *employee* as e, *grad* as g,
    where e.Id = g.Id and Dept = D'

q_4:   call *ta_info* with (D/"CS")

Fig. 18. Simple parameterized views and view queries via call statements

While this approach works fine, it has a problem that we must address. Consider the query *List all teaching assistants and their salaries*. While the Datalog version of this query, i.e., ? ta_info(_, X, _, _, _, _, _, Y), will have no trouble computing the answer, the extended SQL just proposed will fail without further machineries. This is because we now need to call the view ta_info with

    call *ta_info* with ()

without a variable assignment, which breaks down the syntactical structure and possibly raises a runtime exception since the expression Dept = D is not computable. Ideally, we should still be able to use the view and the call should succeed just like in its Datalog counterpart. To achieve this functionality, we introduce the concept of *relational don't care variables* and *relational unification*. We will address these issues in the next section and see how these two concepts become useful in developing the framework of paremeterized views.

Limitations of Macro Expansion and Datalog. The issue of parameter passing to procedures using macro expansion has been a subject of research for some time and such techniques have been utilized in several commercial database systems. It can be argued that macro expansion is powerful enough to handle the functionalities discussed in this section, in the context of the notion of parameterized views and unification discussed shortly. We would like to point out that macro expansion cannot propagate arguments from one view to the next in a way similar to Datalog (sideways information passing). Macro expansion is also static in nature, and is usually handled at compile time (and thus has limited use). In contrast, parameterized views presented here are dynamic in nature and are handled at run time. More importantly, macros, or techniques of macro expansion, cannot be effectively used for knowledge processing and reasoning applications because they require the power of unification, and they are not, in any way, comparable to unification. Freire [13] points out, it is not practical to achieve view parameterization through the integration of deductive engines with SQL, even though an argument for such an approach can be made seemingly naturally.

Then the natural question is, why not choose Datalog? The obvious reason is that almost no production system uses deductive engines such as Datalog as a computational platform, and hence, developing a collaborative model relying on it will not be practical or useful. The technical reason is far more serious even if the hidden web systems had used Datalog as their query processing engines for the following reasons. As discussed in sections 4 and 4.1, the layers of separation in the hidden web implementation requires that we use some form of schema mapping and establish attribute correspondence across the layers. Since schema mapping is inherently related to attribute names (or terms in logic programming terminology), Datalog's name independence will interfere, and it will be extremely difficult, if not impossible, to develop a mapping algorithm that will successfully establish the needed schema correspondences. Not only will Datalog's name independence pose a serious problem, so will its reliance on attribute position for the identification of columns in a relation. So, we believe that parameterized views, as we envision in this paper, is a prudent approach to address the propagation of user constraints to hidden web servers. It is important to note that the parameterized view we introduce here has its own application in relational databases, especially in terms of its capability in reasoning with SQL.

4.3 Syntax and Semantics of Parameterized Views

In this section, we formally define the syntax and semantics of parameterized views in relational databases based on the following principles.

1. We must preserve SQL's tradition of supporting arity independence in accessing a relation, stored (base relations) or derived (views or procedures). This principle is a carryover from traditional practices in SQL.
2. We should be able to use a view at any point in any SQL expression where a relation can be used. Since our goal is to customize views through parameter

passing, it makes sense that we allow parameterized views or procedures in just such a fashion in SQL expressions.
3. We should be able to adorn the variables in the procedure from outside when we reference it in a dynamic fashion at run time.

Keeping these goals in mind, we identify parameterized views as transient views, very much like queries. But unlike queries, they are not evaluated at the point of definition. Instead, they are evaluated every single time they are referenced in other long traditional (long lived) views, queries, or are called using a call statement (to be defined shortly).

We borrow the idea of formal parameter lists in procedural languages albeit uniquely to define the syntax and semantics of parameterized views. We introduce the *with parameter* clause in *create view* statement which defines a maximal set of allowed input parameters for the view. If we treat the attribute list of the select clause within the create view statement as the maximal output list, then the interface to the view in extended SQL is the union of the variables in with parameter and the select clauses as formalized in the definition below.

Definition 1. Let $\mathcal{D}$ be a database, $\mathcal{R}$ be a set of relation names in $\mathcal{D}$, and $\mathcal{A}$ be a set of attribute or tag names. Also let $v \in \mathcal{R}$, $E_1, \ldots, E_l$ be a set of relational expressions involving relations in $\mathcal{D}$, $\{A_1, \ldots, A_n\}$ be a set of expressions involving attributes of $E_1, \ldots, E_l$ and a set of tags $P_1, \ldots, P_m$ such that $\{P_1, \ldots, P_m\} \subseteq \mathcal{A}$. Similarly, let $C_1, \ldots, C_k$ be Boolean conditions involving attributes of $E_1, \ldots, E_l$ and the tags $P_1, \ldots, P_n$. Then,

> create view v with
> parameter $P_1, \ldots, P_m$ as
> select $A_1, \ldots, A_n$
> from $E_1, \ldots, E_l$
> where $C_1 \wedge \ldots \wedge C_k$

is a *parameterized view*. □

The parameter list $P_1, \ldots, P_m$ is used to specify replaceable tags and are being used in the view definition as place holders. The behavior exhibited by the view depends on these place holders, as they play a vital role in the restriction of the computed view. To evaluate the view, we use the *call* statement defined below. The call statement invokes the view with a specified set of bindings for a subset of the tags appearing in the parameter list of the view.

Definition 2. Let $\mathcal{R}$ be a set of relation names, $\mathcal{A}$ be a set of attributes, $v \in \mathcal{R}$ be a parameterized view with a parameter list L, $\{A_1, \ldots, A_n\} \subseteq L \subseteq \mathcal{A}$ be a set of attributes, and finally $c_1, \ldots, c_n$ be a set of constants. Then,

> call v with $(A_1/c_1, \ldots, A_n/c_n)$, and
> call v

are procedure calls. □

Notice that the set $A_1, \ldots, A_n$ in the with clause of the call statement must be an improper subset of the parameter list of the view v. The second form of the call statement allows the possibility of evaluating a view without any parameter passing – i.e., no variable binding, the case where $n = 0$. In this way, it allows us to capture arity polymorphism of the view definition and the parameter list of v becomes a maximal set. This means that we are allowed to use all or a subset of parameters. But we cannot call a parameterized view with an argument that is not a member of its parameter list.

Unfortunately, if $\{A_1, \ldots, A_n\} \subset \{P_1, \ldots, P_m\}$, i.e., a strict subset, then there is a possibility that a run-time error may occur at execution time as the expressions specified in the view may become fragile or non-computable, as demonstrated in the following example. Notice that every view is a relation, and a call to a parameterized view produces a relation by computing the view as a query. Hence, a call statement can be used in an expression where a traditional view or a relation can be used with appropriate renaming as necessary (i.e., as clause). In this way, the *closure* property of declarative views is also preserved.

Consider redefining the ta_info view as follows and calling it with the series of call statements shown next to it. The first call initializes the variables D and I in ta_info with the supplied constants. The computed view is the relation shown in figure 17 (a). Notice that the order of the argument list is immaterial. Current implementation of Oracle supports this form of named assignment of variables in procedure calls when defaults are declared in the procedure for its arguments. Essentially, the argument list in the call statement specifies a set of *substitutions* for the variables in the view expressions. Once the substitutions are made, the query expressed in the select clause can be executed to compute the answer.

v_6:
create view *ta_info* with
parameter *I, D* as
select *Id, Name, Dept, Stipend, Advisor, Salary, Income, Bonus*
from *employee* as e, *grad* as g, call *ta_income* with (D'/D) as i,
  ta_bonus as b
where e.Id = i.Id and i.Id = b.Id and b.Id = g.Id and Dept = D
  and Id = I

q_5: call *ta_info* with $(D/"CS", I/82)$

q_6: call *ta_info* with $(I/82)$

The first procedure call q_5 results in the following query from the parameterized view v_6: ta_info. So, basically the query

q_7:
select *Id, Name, Dept, Stipend, Advisor, Salary, Income, Bonus*
from *employee* as e, *grad* as g, call *ta_income* with (D'/D) as i,
  ta_bonus as b
where e.Id = i.Id and i.Id = b.Id and b.Id = g.Id and Dept = "CS"
  and Id = 82

is executed, but not the view

v_7:
create view ta_info
select $Id, Name, Dept, Stipend, Advisor, Salary, Income, Bonus$
from $employee$ as e, $grad$ as g, call ta_income with (D'/D) as i,
ta_bonus as b
where e.Id = i.Id and i.Id = b.Id and b.Id = g.Id and Dept = "CS"
and Id = 82

as in traditional SQL. This approach makes it really transient and ad hoc, and so the computation does not survive from statement to statement. Hence, a call statement can be used as an expression where a query, view or a relation is used, as noted before.

We, however, have a problem with the second through the fourth call statement above because the substitutions are partial and the where clause in view ta_info becomes fragile due to the presence of non-evaluable conditions, i.e., Dept = D, Id = I, and Dept = D and Id = I, respectively.

There are two ways to deal with this situation. The first is to disallow evaluation of such views at run time. If needed, we may disallow the call itself to avoid generating a run-time error by requiring that the substitution list of the call statement matches the parameter list of the view, i.e., for a view parameter list $P_1, \ldots, P_m$, and a call substitution list $A_1/c_1, \ldots, A_n/c_n$, $m = n$. But this choice restricts the possibility of several interesting ways the call can be perceived.

Another way to deal with this issue is to ignore expressions that do not have a ground substitution. Similar to Datalog, if we take the view that every omitted substitution in the with clause of the call statement is a *"don't care"* term, then we would like to think that the queries

q_8:
select $Id, Name, Dept, Stipend, Advisor, Salary, Income, Bonus$
from $employee$ as e, $grad$ as g, call ta_income with (D'/D) as i,
ta_bonus as b
where e.Id = i.Id and i.Id = b.Id and b.Id = g.Id and Dept = "CS"

and

q_9:
select $Id, Name, Dept, Stipend, Advisor, Salary, Income, Bonus$
from $employee$ as e, $grad$ as g, call ta_income with (D'/D) as i,
ta_bonus as b
where e.Id = i.Id and i.Id = b.Id and b.Id = g.Id and Id = 82

resulting from calls q_4 and q_6 respectively are the likely scenarios in the relational context. While don't care variables are well understood and pose no threat in logic programming paradigm, in relational context, their use in parameterized views, however, raises some interesting questions. In relational model, don't care terms (attributes) are modeled by omitting them in the select clauses. In the context of parameterized views, we are proposing that we treat a predicate involving parameters as don't care terms if certain conditions are met. To guarantee the production of the queries q_8 and q_9 above corresponding to the calls q_4 and q_6, we proceed as follows to define evaluability of expressions and queries.

Definition 3. Let $\mathcal{D}$ be a database, V be a variable name and E be a relational expression over $\mathcal{D}$. Variable V is *non-prime* if V is not an attribute name of a relation in database $\mathcal{D}$. □

Definition 4. Let E be an expression and $\{A_1, \ldots, A_n\}$ be the set of tags appearing in E. Expression E is *evaluable* if E is free of non-prime variables, i.e., E contains only constants or attributes of some relations in the database. An SQL query is evaluable if all expressions in select and where clauses are evaluable. □

Our goal is to allow parameterized views of the form ta_info yet guarantee query safety, and thus ensure that every expression in a query is evaluable at run time even when a subset of parameters are used in the call statement. To guarantee query safety, we define relational don't care expressions as follows.

Definition 5. Let Q be an SQL query and E be an expression in the select or where clause of Q. Then E is a *relational don't care* expression if E is not evaluable. □

During execution, don't care expressions in select clause are treated as null, and as true in where clause. Hence, it is safe to remove don't care expressions from the select list, and from the where clause. Under these assumptions, queries q_4 and q_6 on view v_6 would result in the execution of the queries q_8 and q_9 as intended. Note that the introduction of the concept of relational don't care expressions gives rise to the following two definitions of safe and unsafe queries and of intended definitions of parameterized views.

Definition 6. Let v be a parameterized view, and L be its parameter list. Let C be a call statement for v with a substitution list S. If T is the set of tags in S then, C is *safe* if $T \subseteq L$, otherwise it is *unsafe*. □

Definition 7. Let v be a parameterized view, and L be its parameter list. Let D be the set of queries obtained from v by replacing expressions in v involving elements $l \in L$ in all possible ways (there are exactly $|\mathcal{P}(L)|$ possible ways) as follows:

- replace an expression involving l in the select clause with null.
- replace an expression involving l in the where clause with true.
- replace an expression involving l in the from clause with empty.

Then, the set D is the *intended definitions* of v. □

4.4 Relational Unification Algorithm

Before we proceed to develop a relational unification algorithm to support processing parameterized views, we need to make a distinction between variables that are external to an expression as opposed to those that are internal. Because our goal was to support arity independence following the tradition of SQL, we need to develop a method to simulate name independence so that names in other expressions do not conflict with the variable names in an expression at hand. This will help control accidental instantiation of variables across expressions, and alleviate concern about name duplication in other expressions.

Definition 8. Let E be a relational expression, and V be the set of variables appearing in E. A variable $var \in V$ is *external* to E if it is in a substitution list of a call statement in the form var/e, it is *internal* otherwise. □

The essence of definition 8 is that a variable (i.e., var) that refers to a parameter of a parameterized view in a substitution list of a call statement embedded in an expression E is independent of the expression E. So, any possible substitution in E must not effect var. Recall the view ta_info defined in example 1. In that definition, we needed to separate the variable D in ta_info with the variable D' in view ta_income. Otherwise, any uniform replacement of D in ta_info would effect ta_income, or may even result in meaningless substitutions (i.e., as in this case "CS"/"CS"). Now, we can use the idea of external variables to our advantage as shown in the example 2 below, yet support name independence from expression to expression. The unification algorithm we present in this section utilizes this notion.

---

algorithm: unification
input: a parameterized view v and a substitution list L.
output: a select query, q.
begin
   let $\{A_1, \ldots, A_k\}$ be the set of external variables in L;
   let $\{P_1, \ldots, P_n\}$ be the set of parameters in v;
   if $\{A_1, \ldots, A_k\} \not\subseteq \{P_1, \ldots, P_n\}$ then
      report runtime unification error;
      exit;
   apply substitution uniformly on the <u>internal</u> variables of view v to obtain v' as
      $v' = v[\{A_1/c_1, \ldots, A_k/c_k\}]$;
   extract select query from v' as q;
   for every select clause expression e of q
      if e is not evaluable then
         remove e from q;
   for every where clause condition expression c of q
      if c is not evaluable then
         replace c with Boolean constant `true`;
   return q;
end.

Fig. 19. Relational unification algorithm

Example 2. Consider the views v_4 and v_6 and the query q_4. The algorithm for relational unification presented in figure 19 will produce the query q_8 on both views v_4 and v_6. Notice that variable D in the subexpression call ta_income with $(D/"CS")$ as i is untouched as it is external to this expression and the fact that view ta_income also has a variable name D is not an issue. In logic programming, this is a by product of unification and substitution. □

The goal of relational unification is somewhat different from its logic programming counterpart. In logic programming, the goal is to make the target predicate (consequent of a rule, and thus the body) look as identical as possible to the query predicate. The safety of the evaluation is not an issue. But in relational unification, we make it a point that the target query remains safe, evaluable and intended as defined in definitions 6, 4 and 7 respectively. For this, we rely on the parallel that exists in logic programming, and hence the approach remains grounded on sound foundations. For any extended SQL call statement C, and a parameterized view v, algorithm unification always returns an evaluable query which is intended if, and only if, C is safe and relational unifiable, or else it returns an error message.

5 Constraint Injection into Hidden Web Form Views

We are now equipped to formally define the concept of *constraint injection* to SQL views. We will utilize the fact that web forms already accept user constraints as a set of attribute-value pairs that they plug into view definitions for onward evaluation, and that we now have a mechanism to influence the restriction operation or the selection condition of views through parameterization. We formally define the concept of constraint injection as follows.

Definition 9. Let W be a hidden web form, and V_w be the set of form variables appearing in W. Let V be the parameterized view to compute the query response R such that V_w appears in the Boolean condition of V as C_{V_w}. Let V_r be the set of attributes in the computed view R by V. Then, V is a *fully parameterized view* if the parameter list L of V is equal to $V_w \cup V_r$. □

We assume that applications are already aware of the form variables, and anticipate finding a table in response to the form submission that will have an associated scheme. The application also expects to apply a selection condition θ and project out the unwanted columns. In other words, for a relation r, we have a user supplemental query Q of the form

    select $A_1, \ldots, A_n$
    from r
    where θ

on the view r. Our goal is to transmit this query to W and enforce θ as an additional constraint on a fully parameterized view V and apply projection $\Pi_{A_1,\ldots,A_n}(r)$. Since the view V is still a black-box, we will apply our best effort model to define an evaluable parameterized view as follows.

Definition 10. Let V be a fully parameterized view with defined constraint C, a formal parameter list L, and a select list with L_v. Let Q be a supplemental query on V with projection list P and constraint θ. A *constrained view* V_c is a rewriting of V such that the select list L_v is replaced with P and the constraint C is replaced with $C \wedge (\theta)$. □

Although we have modified the view to behave as though the original view V has been computed as a table r, and we are now applying the supplemental query on

r, we still need to guarantee that the rewriting is syntactically and semantically correct. This is because the project list P may not be appropriate for V or the condition θ may not be enforceable because of the schema mismatch. To ensure that we have a best effort view that is computable at site W, we define an admissible view as follows.

Definition 11. Let V_c be a constrained view of a fully parameterized view V, Q be a supplemental query, μ be a binary mapping function that uniquely maps a set of terms to another set of terms, and V_q be the set of all attribute names in the select list of Q and condition θ. V_c is *strictly admissible* if $\forall t, t \in V_q \rightarrow (\exists u, u \in V_r \wedge \mu(t, u) \neq 0)$. It is *best effort admissible* if $\forall t, t \in V_q \rightarrow (\exists u, u \in V_r \wedge \mu(t, u) \neq 0)$ holds, or t (or condition involving t in θ) is removed from the project list P (or boolean term involving t is replaced with the constant true in θ) otherwise. □

Definition 11 makes it possible to either enforce all requirements of supplemental query Q (strict admissibility), or best effort enforcement by removing the projections and selection conditions that are not enforceable. By changing a condition to true for terms that are not computable (because the attribute does not exist in the view), we retain the logical structure of θ unaltered. Once the schema mapping is completed, and we know if we are able to enforce strict or best effort admissibility, we are in a position to honor application request, or reject it. All an application now has to do is send the web form entries as a pair of attributes and values, a project list, and selection condition and a choice of mapping function to the site W. Since we apply θ as a conjunct, we can never violate site security, as θ is in addition to what V was already applying (an instance of the case $\sigma_a(\sigma_b(r)) \equiv \sigma_{a \wedge b}(r)$). Finally, the technology we have proposed for parameterized views handles the propagation of these constraints in θ, called *constraint injection*, deeper into the view definition appropriately.

6 Summary

Our central focus in this paper was to propose a model of collaboration for Life Sciences databases to improve overall computational efficiency. By introducing the notion of parameterized views and then admissible constrained views, we have demonstrated that user constraints can now be enforced on the server side view computation without any security breach or SQL injection. We emphasize that parameterized views can be used as a substitute for hidden web interface implementation, a form of module construct to avoid string substitution based interface query rewriting that initially led to SQL injection breach in traditional systems. We believe that the techniques presented in this paper, though simple, for the first time make it possible for remote sites to accept user directed computation inside the hidden web in a secure and collaborative manner without the risk of SQL injection. A full treatment of a more formal model for constraint injection was outside the scope of this paper due to space constraint.

A more detailed discussion on this issue may be found in [22] where we introduce the idea of mandatory and optional parameters and avoid using relational don't care variables (as we did in this paper), and introduce the idea of *interface views*. Constraint injection is given a formal status by defining a simple notion of *query containment* along the line of earlier research in [1,12,9,14,25]. The presented model respects site autonomy and improves overall query efficiency with site participation. What remains to be explored is how the coordination plan can be developed, transmitted to participating sites and implement the overall architecture which we plan to address in a separate paper. We see opportunities to utilize existing research on distributed transaction processing, and commit protocols in database literature toward developing query plans as part of our future research endeavors.

References

1. Afrati, F.N., Damigos, M., Gergatsoulis, M.: Query containment under bag and bag-set semantics. Inf. Process. Lett. 110(10), 360–369 (2010)
2. Altintas, I., Berkley, C., Jaeger, E., Jones, M., Ludascher, B., Mock, S.: Kepler: An extensible system for design and execution of scientific workflows. In: SSDBM, p. 423 (2004)
3. Bancilhon, F., Maier, D., Sagiv, Y., Ullman, J.D.: Magic sets and other strange ways to implement logic programs. In: PODS, pp. 1–15 (1986)
4. Benson, D.A., Karsch-Mizrachi, I., Lipman, D.J., Ostell, J., Wheeler, D.L.: Genbank. Nucleic Acids Res. 36(database issue) (January 2008)
5. Berman, H.M., Westbrook, J., Feng, Z., Gilliland, G., Bhat, T.N., Weissig, H., Shindyalov, I.N., Bourne, P.E.: The protein data bank. Nucleic Acids Res. 28(1), 235–242 (2000)
6. Bhattacharjee, A., Islam, A., Amin, M.S., Hossain, S., Hosain, S., Jamil, H., Lipovich, L.: On-the-fly integration and ad hoc querying of life sciences databases using LifeDB. In: 20th International Conference on Database and Expert Systems Applications, Linz, Austria, pp. 561–575 (August 2009)
7. Boulakia, S.C., Biton, O., Davidson, S.B., Froidevaux, C.: Bioguidesrs: querying multiple sources with a user-centric perspective. Bioinformatics 23(10), 1301–1303 (2007)
8. Cafarella, M.J., Halevy, A.Y., Khoussainova, N.: Data integration for the relational web. PVLDB 2(1), 1090–1101 (2009)
9. Calvanese, D., Giacomo, G.D., Lenzerini, M., Vardi, M.Y.: View-based query containment. In: PODS, pp. 56–67 (2003)
10. Ceri, S., Gottlob, G., Tanca, L.: What you always wanted to know about datalog (and never dared to ask). IEEE Trans. Knowl. Data Eng. 1(1), 146–166 (1989)
11. Chen, L., Jamil, H.M.: On using remote user defined functions as wrappers for biological database interoperability. International Journal on Cooperative Information Systems 12(2), 161–195 (2003)
12. Farré, C., Teniente, E., Urpí, T.: Checking query containment with the cqc method. Data Knowl. Eng. 53(2), 163–223 (2005)
13. Freire, J.: Practical problems in coupling deductive engines with relational databases. In: Proceedings of the 5th KRDB Workshop, Seattle, WA, pp. 11-1–11-7 (May 1998)

14. Grahne, G., Thomo, A.: Query containment and rewriting using views for regular path queries under constraints. In: PODS, pp. 111–122 (2003)
15. Guo, S., Dong, X., Srivastava, D., Zajac, R.: Record linkage with uniqueness constraints and erroneous values. PVLDB 3(1), 417–428 (2010)
16. Gusfield, D., Stoye, J.: Relationships between p63 binding, dna sequence, transcription activity, and biological function in human cells. Mol. Cell 24(4), 593–602 (2006)
17. He, B., Zhang, Z., Chang, K.C.C.: MetaQuerier: querying structured web sources on-the-fly. In: SIGMOD Conference, pp. 927–929 (2005)
18. Hosain, S., Jamil, H.: An algebraic foundation for semantic data integration on the hidden web. In: Third IEEE International Conference on Semantic Computing, Berkeley, CA (September 2009)
19. Hossain, S., Jamil, H.: A visual interface for on-the-fly biological database integration and workflow design using VizBuilder. In: 6th International Workshop on Data Integration in the Life Sciences (July 2009)
20. Hull, D., Wolstencroft, K., Stevens, R., Goble, C., Pocock, M.R., Li, P., Oinn, T.: Taverna: a tool for building and running workflows of services. Nucleic Acids Res. 34 (July 2006); web Server issue
21. Jamil, H., Islam, A., Hossain, S.: A declarative language and toolkit for scientific workflow implementation and execution. International Journal of Business Process Integration and Management 5(1), 3–17 (2010); iEEE SCC/SWF 2009 Special Issue on Scientific Workflows
22. Jamil, H., Jagadish, H.V.: Accepting external constraints on deep web database query forms and surviving it. Tech. rep., Department of Computer Science, Wayne State University, Michigan (June 2011)
23. Kent, J.W., Sugnet, C.W., Furey, T.S., Roskin, K.M., Pringle, T.H., Zahler, A.M., Haussler, D.: The human genome browser at ucsc. Genome Res. 12(6), 996–1006 (2002)
24. Köpcke, H., Thor, A., Rahm, E.: Evaluation of entity resolution approaches on real-world match problems. PVLDB 3(1), 484–493 (2010)
25. Penabad, M.R., Brisaboa, N.R., Hernández, H.J., Paramá, J.R.: A general procedure to check conjunctive query containment. Acta Inf. 38(7), 489–529 (2002)
26. Roichman, A., Gudes, E.: Fine-grained access control to web databases. In: SACMAT, pp. 31–40 (2007)
27. Sismanis, Y., Brown, P., Haas, P.J., Reinwald, B.: GORDIAN: efficient and scalable discovery of composite keys. In: VLDB 2006, pp. 691–702 (2006)
28. Tejada, S., Knoblock, C.A., Minton, S.: Learning object identification rules for information integration. Inf. Syst. 26(8), 607–633 (2001)
29. Wang, K., Tarczy-Hornoch, P., Shaker, R., Mork, P., Brinkley, J.: Biomediator data integration: Beyond genomics to neuroscience data. In: AMIA Annu. Symp. Proc., pp. 779–783 (2005)
30. Yakout, M., Elmagarmid, A.K., Elmeleegy, H., Ouzzani, M., Qi, A.: Behavior based record linkage. PVLDB 3(1), 439–448 (2010)

Flexible-ICA Algorithm for a Reliable Iris Recognition

Imen Bouraoui[1], Salim Chitroub[1], and Ahmed Bouridane[2,3]

[1] Signal and Image Processing Laboratory, Electronics and Computer Science Faculty, USTHB,
P.O. Box 32, El – Alia, Bab – Ezzouar, 16111, Algiers, Algeria
[2] School of Computing, Engineering and Information Sciences, Northumbria University,
Pandon Building Newcastle upon Tyne, UK
[3] Department of Computer Science, King Saud University, P.O. Box 2454,
Riyadh, 11451, Saudi Arabia
imbouraoui@yahoo.fr, s_chitroub@hotmail.com
ahmed.bouridane@unn.ac.uk, abouridane@ksu.edu.sa

Abstract. In many large scale biometric-based recognition problems, knowledge of the limiting capabilities of underlying recognition systems is constrained by a variety of factors including a choice of a source encoding technique, quality, complexity and variability of collected data. In this paper, we propose a novel iris recognition system based-on Independent Component Analysis (ICA) encoding technique, which captures both the second and higher-order statistics and projects the input data onto the basis vectors that are as statistically independent as possible. We apply Flexible-ICA algorithm in the framework of the natural gradient to extract efficient feature vectors by minimizing the mutual information of the output data. The experimental results carried on two different subsets of CASIA V3 iris database show that ICA reduces the processing time and the feature vector length. In addition, ICA has shown an encouraging performance which is comparable to the best iris recognition algorithms found in the literature.

Keywords: Biometrics, Iris recognition, Feature extraction, Flexible-ICA, CASIA-V3 iris database.

1 Introduction

In recent years, with the increasing demands of security in our networked society, technologies for personal authentication are becoming one of the main solutions to safeguard people's privacy and properties. Traditional user authentication schemes are based on passwords, secret codes and/or identification cards or tokens, can be cracked by intercepting the presentation of such a password, or even by counterfeiting it (via passwords dictionaries or, in some systems, via brute force attacks). On the other hand, an intruder can attack systems based on identification card or tokens by robbing, copying or simulating them. In this case, biometric technology becomes an attractive alternative recognition technique [1]. Biometric authentication deals with recognizing the identity of individuals based on their unique physical or behavioural characteristics. Physical characteristics such as fingerprint, palm print, hand geometry, face, ear, voice, and iris patterns or behavioural attributes such as gait, typing pattern and handwritten signature present information that is specific to the person and can be used in the authentication process [2]. Out of all these biometric features, fingerprint verification has received considerable attention and has been successfully used in law enforcement applications. Face recognition and speaker recognition have also been widely studied over the last years, whereas iris recognition is a newly emerging approach to personal identification [3], [4]. It is reported in [5] that iris recognition is one of the most reliable biometrics.

The iris is an internal organ of the eye that is located just behind the cornea and in front of the lens. The functionality of the iris is to control the size of the pupil, which in turn regulates the amount of light entering the pupil and impinging on the retina. The visible structures around the iris are the pupil, sclera and cornea. The pupil lies near the centre of the iris. It appears dark because most of the light entering the pupil is absorbed by the tissues inside it. The sclera is commonly known as "the white of the eye". It is the eye's protective outer cover. The cornea is the transparent part of the human eye which covers the iris, pupil and the anterior part of the human eye [35]. The iris is an annulus structure consisting of fibro-vascular tissues called stroma. It is divided into two regions; pupillary zone which is the inner region of the iris on the periphery of the pupil and the ciliary zone which constitutes the rest of the iris as shown in Figure 1. These two regions are separated by the collarette which typically represents the thickest part of the human iris. Furrows found in the stroma on the circumference of the iris have a circular structure. They are termed as circular furrows. The majority of the furrows in the stroma is interlaced and radiate towards the pupil (dilator muscles) and are termed as radial furrows. Some circular furrows (sphincter muscles) are present just outside the pupil in a narrow band of about 1 mm. The stroma connects the sphincter muscles, which contract the pupil, with the dilator muscles, which dilate the pupil. The iris also contains sharply demarcated crypts that are a result of iris thinning which exposes the darkly pigmented posterior layer. Thus, the iris in the presence of near-infrared lighting, is observed to have several features including radial and concentric furrows, crypts and the collarette, all of which contribute to its uniqueness and play a significant role in recognising an individual.

Ophthalmologists originally proposed that the iris of the eye might be used as a kind of optical fingerprint for personal identification [6]. Their proposal was based on clinical results that every iris is unique and it remains unchanged in clinical photographs. The human iris begins to form during the third month of gestation. The structure is complete by the eighth month of gestation, but pigmentation continues into the first year after birth. It has been discovered that every iris is unique and no two people even two identical twins have uncorrelated iris patterns [7], and is stable throughout the human life. It is suggested in recent years that the human irises might be as distinct as fingerprint for different individuals, leading to the idea that iris patterns may contain unique identification features. However, there are various medical conditions that may affect iris [36], [37], [38], [39], such as cataract, glaucoma, albinism, etc.

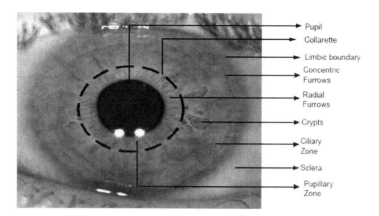

Fig. 1. Anatomy of iris. The two white blobs in the pupil are the specular refections due to the imaging device.

Many researchers have worked on iris recognition including image databases, and human iris authentication process basically consists of four steps as follows: (a) iris segmentation, where the iris is localized and isolated from the noise due to sclera, pupil, eyelids and eyelashes; (b) normalization, where iris is mapped from rectangular representation to domain polar representation; (c) feature extraction, where a feature vector is formed which consists of the ordered sequence of the features extracted from the various representation of the iris images; (d) and matching, where the feature vectors are classified through different techniques such as Hamming Distance, weight vector and winner selection, dissimilarity function, etc.

In our work, we first use Canny edge detection and Hough transform for iris localisation. Then, the extracted iris region is normalised into a rectangular block with constant dimensions to account for imaging inconsistencies of Daugman's model. We apply Flexible-ICA algorithm to extract efficient feature vectors. Then, each iris feature vector is encoded into an iris code. Finally, a Hamming distance is used for the matching process. We demonstrate our experimental results using two different subsets of CASIA-V3 iris image database and some mathematical criteria, in order to compare our technique against some other existing methods in order to assess its usefulness.

The remainder of this paper is organized as follows. In Section 2 we briefly review some existing and important iris recognition systems. Section 3 is concerned with description of iris image pre-processing, which involves iris localization, and normalization. Basic principles of iris feature extraction using flexible-ICA algorithm are reviewed in detail in Section 4. Section 5 introduces matching technique. Section 6 reports the experiments and results. And finally, a conclusion and future work are given in Section 7.

2 Overview of Some Iris Recognition Systems

In this section we briefly review a sample of iris recognition works available in the literature with respect to the basic constituting modules, namely: iris preprocessing including location and normalisation, feature extraction and encoding, and matching.

The idea of automating iris recognition was proposed by Flom and Safir, in 1987. They obtained a patent for an unimplemented conceptual design of an automated iris biometric recognitions system [8]. Their description suggested highly controlled conditions, including a headrest, a target image to direct the subject's gaze, and a manual operator. To account for the expansion and contraction of the pupil, they suggested changing the illumination to force the pupil to a predetermined size. While the imaging conditions that they describe may not be practical, some of their other suggestions have clearly influenced later research. For example, they suggested using pattern recognition tools, including difference operators, edge detection algorithms, and the Hough transform, to extract iris descriptors. To detect the pupil, they suggest an algorithm that finds large connected regions of pixels with intensity values below a given threshold. They also suggested that a description of an individual's iris could be stored on a credit card or identification card to support a verification task.

One of the well-known and thoroughly tested algorithms is due to Daugman [9], [19]. The Daugman algorithm first locates the pupillary and limbic boundaries of the iris using an integro-differential operator that finds the circles in the image where the intensity changes most rapidly with respect to changes in the radius. Once located, the iris image is converted from Cartesian form by projecting it to onto a dimensionless pseudo-polar coordinate system. The iris features are encoded and a signature is created using a 2-D complex-valued Gabor filter, where the real and imaginary parts of each outcome are assigned a value of 0 or 1 according to whether they are negative or positive, i.e. only the quadrant of the phase is encoded. Finally, two images are said to be independent if their fractional Hamming distance is above a certain threshold. Otherwise they are match with the Hamming distance being equals to the number of mismatching bits divided by the number of compared bits. The Daugman algorithm has been essentially error-free when applied to a very large database [9].

The Wildes algorithm [10] locates the iris boundaries by creating a binary edge map using gradient-based edge detection, and then finds the centers and radii of these circles using a Hough Transform. The upper and lower eyelids are located similarly using parabolic arcs. Rather than map every iris image to a common system of polar coordinates, the Wildes algorithm compares two images by geometrically warping one image, via shifting and rotations, until it is a best fit with the other image, in the sense of minimizing the mean square distance. A Laplacian pyramid is constructed at four different resolution levels to encode the image data. Matching is achieved via an application of normalized correlation and Fisher's linear discriminant [10].

Boles and Boashash [11] have given an algorithm that locates the pupil center using an edge detection method, records grey level values on virtual concentric circles, and then constructs the zero-crossing representation on these virtual circles based on a one-dimensional dyadic wavelet transform. Corresponding virtual circles in different images are determined by rescaling the images to have a common iris diameter. The authors create two dissimilarity functions for the purposes of matching, one using every point of the representation and the other using only the zero crossing points. The algorithm has been tested successfully on a small database of iris images, with and without noise [11].

In [12], Lim, et al. propose alternative approaches to both feature extraction and matching, following a standard iris localization and conversion to polar coordinates relative to the center of the pupil. For feature extraction they compare the use of the Gabor Transform and the Haar Wavelet Transform, and their results indicate that the Haar Transform somewhat performs better. Using the Haar transform the iris patterns can be stored using only 87 bits, which compares well to the 2,048 required by Daugman's algorithm, for example. The matching process uses an LVQ competitive learning neural network, which is optimized by a careful selection of initial weight vectors. Also, a new multidimensional algorithm for winner selection is proposed. Experimental results are given in [12] based on a database of images of irises from 200 people.

In [13], Tisse, et al., present a modification of Daugman's algorithm, with two major differences. The two innovations relate to the iris location and feature extraction stages. The use of a dimensionless polar coordinates and Hamming distance remain the same. To locate the iris, the Tisse algorithm applies a gradient decomposed Hough Transform to find the approximate center of the pupil, and then applies an integro-differential operator similar to Daugman's algorithm, to find the precise locations of the iris boundaries. This combined approach has the advantage of avoiding errors due to specular reflection in the images. In the feature extraction and encoding step, Hilbert Transform is used to create an analytic image, whose output is then encoded as an emergent frequency vector and an instantaneous phase. This approach has an advantage of being computationally efficient. The Tisse algorithm has reportedly been successful when tested on a database of real iris images [13].

Tan, et al., suggest several innovations of iris recognition in [14], and [15], and then provide a comparison of different methods and algorithms. The iris is localized in several steps which first find a good approximation for the pupil center and radius, and then apply the Canny operator and the Hough transform to locate the iris boundaries more precisely. The iris image is converted to dimensionless polar coordinates, similarly to Daugman, and then is processed using a variant of the Gabor filter. The dimension of the signature is reduced via an application of the Fisher linear discrimant. The L1 distance, L2 distance (i.e. Euclidean distance), and cosine similarity measures are considered for matching. A careful statistical performance evaluation is provided for the authors' work, and for most of the well-known algorithms mentioned above.

In another approach presented by Ma et al., the quality of image is assessed with the help of Support Vector Machines [15]. The detection is done using Hough Transform and the feature vectors are generated using a set of Multichannel Spatial Filters and Even Symmetry Gabor filters to capture local texture information of the iris, which is then used to construct a fixed length feature vector. The dimensionality reduction of the feature vectors is done using Fisher Linear Discriminant. Nearest feature line method is used for iris matching. In [16] a set of one-dimensional intensity signals is constructed to effectively characterize the most important information of the original two-dimensional image using a particular class of wavelets; a position sequence of local sharp variation points in such signals is recorded as features. A fast matching scheme based on exclusive OR operation is used to compute the similarity between a pair of position sequences.

In [17], several interesting ideas can also be found in the work of Du et al.. An edge detection is performed using the Canny method, and each iris image is then converted into standardized polar coordinates relative to the center of the pupil. The feature extraction stage is quite different from those mentioned previously, and is simple to implement. The authors use a gray scale invariant called Local Texture Patterns (LTP) that compares the intensity of a single pixel to the average intensity over a small surrounding rectangle. The LTP is averaged in a specific way to produce the elements of a rotation invariant vector. Thus the method performs a lossy projection from 2D to 1D. This vector is then normalized so that its elements sum to 1. The matching algorithm uses the "Du measure", which is the product of two measures, one based on the tangent of the angle between two vectors p and q, and the other based on the relative entropy of q with respect to p, otherwise known as the Kullback-Liebler distance. Another paper involving Du [18], in the context of hyperspectral imaging, provides evidence that the Du measure is more sensitive than either of the other two measures. This iris recognition algorithm is quite fast and appears to be most appropriate for producing a "watchlist" [17], rather than being used for identification of specific individuals.

In [30], Meena used high thresholding technique for pupil detection and major intensity change for iris detection. She has developed five different algorithms for iris recognition based on circular and radial features, Fourier transform along radial direction, circular-mellin filters, corner detection and local binary patterns. She has also reported the use of the fusion of circular-mellin and corner detection algorithm to extract features for iris pattern recognition. Vatsa et al. [31] have used thresholding and Freeman's chain code algorithm for detecting pupil and also used linear contrast filter for detecting iris. They have used 1-D log polar Gabor wavelet and Euler numbers to extract texture features for iris pattern recognition. Monro et al. [32] used edge based segmentation for pupillary detection. They also have scanned the horizontal line through the pupil center and found jumps in gray level on either side of the pupil for iris boundary detection. They have developed an iris feature extraction method based on discrete cosine transform (DCT). They applied the DCT to overlapping rectangular image patches rotated 45 degrees from the radial axis. The differences between the DCT coefficients of adjacent patch vectors are then calculated and a binary code is generated from their crossings. In order to increase the speed of the matching, three most discriminating binarized DCT coefficients are kept, and the remaining coefficients are discarded.

Few approaches based on the emergent method "Independent Component Analysis (ICA)" are also proposed in [1], [33] and [34]. Dorairaj et al. [33] used a combination of Principal Component Analysis with Independent Component Analysis to encode the iris image. They used the integro-differential operators to localize the iris and the pupil. The matching was accomplished using both the Euclidean and the Hamming distance metrics. In [1] Wang et al. proposed a frame of iris recognition, they used multi-scale approach to realize iris localization and a method to represent iris feature with ICA including fixed-point algorithm which is an approximation of Fast-ICA algorithm [40]. They obtained a correct recognition rate of 97.25%, and the computational load for calculating feature vectors was very high. However, results obtained by Bae et al. [34], using a database of 990 images including 99 subjects, show that Fast-ICA algorithm enhance the performance of iris feature extraction which is 0.113 %, and reduced the computational complexity and feature vector size to 22.9 ms and 49 Bytes, respectively.

Although significant progress has been achieved in iris recognition, some problems remain unsolved. Most of developed systems and algorithms are claimed to have exclusively high performance and generally used small image sets for performance evaluation. In our work, we describe and analyze the performance of flexible-ICA algorithm, new iris feature extraction technique, by partitioning images into the patches and reducing their dimensionality. Two different, large and real subsets of international certified CASIA iris image databases are used for testing our implemented method. We compare our results with the results obtained in [9], [15] and [16]; they used larger images set involving more than 200 subjects.

3 Image Pre-processing

The iris is an annular part between the pupil (inner boundary) and the sclera (outer boundary). Therefore, a captured image cannot be expected to have only the iris part, it contains some non-useful part e.g. sclera, eyelid and pupil, therefore the iris region should be located in captured eye image, and normalized to polar array.

3.1 Iris Localisation

Iris localization by definition means to isolate the actual iris region in a digital eye image by detecting the inner and outer boundaries of the iris. The eyelids and eyelashes normally occlude the upper and lower parts of the iris region. To detect the iris and pupil boundaries, Hough transform is used by involving Canny edge detection to generate an edge map. The gradients are biased in the vertical direction for the outer iris/sclera boundary while the vertical and horizontal ones are weighted equally for the inner iris/pupil boundary, as suggested in [13] and [10].

The Hough transform locates contours in an n-dimensional parameter space by examining whether they lie on curves of a specified shape. For the iris outer or pupillary boundaries and a set of recovered edge points (x_i, y_i), $i = 1, \ldots, n$, a Hough transform is defined by

$$H(x_c, y_c, r) = \sum_{i=1}^{n} h(x_i, y_i, x_c, y_c, r) \qquad (1)$$

where $H(x_c, y_c, r)$ shows a circle through a point, the coordinates x_c, y_c, r define a circle by the following equation

$$x_c^2 + y_c^2 - r^2 = 0 \qquad (2)$$

In the case of edge detection for iris boundaries, the above equation becomes

$$(x_i - x_c)^2 + (y_i - y_c)^2 - r^2 = 0 \qquad (3)$$

The eyelids are then isolated by first fitting a line to the upper and lower eyelid parts using a linear Hough transform. A second horizontal line is then drawn, which intersects with the first line at the iris edge that is closest to the pupil. The second horizontal line allows a maximum isolation of eyelid regions while a thresholding operation is used to isolate the eyelashes.

3.2 Iris Normalisation

Normalization refers to *preparing* a localised iris image for the feature extraction process. The process involves unwrapping the iris image and converting it into its polar equivalent. It is carried out by using Daugman's Rubber sheet model [22], [19] as shown in figure 2. The centre of the pupil is considered as the reference point and a remapping formula is used to convert the points on the Cartesian scale to the polar scale.

The remapping of iris image $I(x, y)$ from raw Cartesian coordinates to polar coordinates (r, θ) can be represented as

$$I(x(r,\theta), y(r,\theta)) \rightarrow I(r,\theta) \qquad (4)$$

where r is on the interval $[0,1]$ and θ is angle $[0, 2\pi]$, with

$$\begin{cases} x(r,\theta) = (1-r)x_P(\theta) + rx_I(\theta) \\ y(r,\theta) = (1-r)y_P(\theta) + ry_I(\theta) \end{cases} \qquad (5)$$

where

$$\begin{cases} x_P(\theta) = Ox_P(\theta) + r_P \cos(\theta) \\ y_P(\theta) = Oy_P(\theta) + r_P \sin(\theta) \end{cases} \qquad (6)$$

and

$$\begin{cases} x_I(\theta) = Ox_I(\theta) + r_I \cos(\theta) \\ y_I(\theta) = Oy_I(\theta) + r_I \sin(\theta) \end{cases} \qquad (7)$$

The centre of the pupil is denoted by (Ox_P, Oy_P) and (Ox_I, Oy_I) is the center of the iris; r_P is the radius of the pupil and r_I is the radius of the iris; and (x_P, y_P) and (x_I, y_I) are the coordinates of points bordering the pupil's radius and iris' radius respectively along the direction θ.

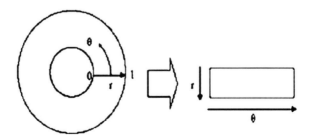

Fig. 2. Daugman's Rubber sheet model

4 Feature Extraction by ICA

The iris has an interesting structure and presents rich texture information. The distinctive spatial characteristics of the human iris are available at a variety of scales [7]. As such, a well-known subspace analysis technique such as Independent Component Analysis (ICA) is used to capture local distinctive information in an iris and creates a set of compact features for an effective recognition task.

4.1 Independent Component Analysis

ICA represents a novel and powerful statistical method for subspace analysis, with applications in computational neuroscience and engineering. It consists of automatically identifying the underlying components in a given data set. It requires that at least as many simultaneously recorded mixtures as there are components and each mixture is a combination of components that are independent and nongaussian. However, like all methods, the success of ICA in a given application depends on the validity of the assumptions on which ICA is based and the results should be treated with caution. So, much theoretical work remains to be done on precisely how ICA fails when its assumptions, i.e. linear mixing and statistical independence, are severely violated [23], [20].

Generally, the most popular noising – free linear model of ICA is expressed as follows

$$X = AS \qquad (8)$$

where X is a vector variable, of dimension N, in which each variable is an observed signal mixture and S is a vector variable, of dimension M, in which each variable is a source signal. We assume that $N \geq M$. The mixing matrix A defines a linear transformation on S, which can usually be reversed in order to recover an estimate U of S from X, i.e.

$$S \approx y = WX \qquad (9)$$

where the separating matrix $W = A^{-1}$ is the inverse of A. However, A is an unknown matrix and cannot therefore be used to find W. Instead, many iterative algorithms are used to approximate W in order to optimize independence of S. In this paper, the Flexible-ICA algorithm [24] is deployed.

Since mutual information is the natural information-theoretic measure of the independence of random variables, it could be used as the criterion for finding the ICA transform. In this approach, which is an alternative to the model estimation approach, the ICA of a random vector X is defined as an invertible transformation as in (9), where the matrix W is determined so that the mutual information of the transformed components of S is minimized.

Mutual information is a natural measure of the dependence between random variables. It can be interpreted by using the concept of differential entropy H of a random vector y with density $f(.)$ as follows [26]

$$H(y) = -\int f(y) \log f(y) dy \tag{10}$$

Entropy is considered as the coding length of the random variable y_i, $i = 1...N$. In fact, it is defined as

$$H(y_i) = -\sum_i P(y_i) \log P(y_i) \tag{11}$$

However, mutual information I between the N (scalar) random variables y_i, $i = 1...N$ [25], [27], is defined as

$$I(y_1, y_2, ..., y_N) = \sum_{i=1}^{N} H(y_i) - H(y) \tag{12}$$

Using the invertible linear transformation presented in (9). Mutual information [25], [26], is given by

$$I(y_1, y_2, ..., y_N) = \sum_i H(y_i) - H(x) - \log|\det W| \tag{13}$$

To search space of separating matrix or *Stiefel manifold* W, let us consider that y_i have been *uncorrelated* and have unit variance. This means

$$E[yy^T] = WE[xx^T]W^T = I \tag{14}$$

which implies

$$\begin{aligned} \det I &= 1 \\ &= \det WE[xx^T]W^T \\ &= \det W \det E[xx^T] \det W^T \end{aligned} \tag{15}$$

This requires that (det$W$) must be constant. In this case, the minimisation of mutual information leads to the following loss function

$$L(W) = -\sum_{i=1}^{N} \log p_i(y_i) \tag{16}$$

The gradient of loss function (16) is given by

$$\nabla L(W) = \frac{\partial L(W)}{\partial W} = \varphi(y)x^T \tag{17}$$

where

$$\varphi(y) = [\varphi_1(y_1), ..., \varphi_N(y_N)]^T \tag{18}$$

and

$$\varphi(y_i) = -\frac{d \log p_i(y_i)}{dy_i} \quad (19)$$

The natural Reimannian gradient in Stiefel Manifold was calculated by [28] and it can be written as follows

$$\begin{aligned}\overline{\nabla}L(W) &= \nabla L(W) - W[\nabla L(W)]^T W \\ &= \varphi(y)x^T - y\varphi^T(y)W \end{aligned} \quad (20)$$

With this, the learning algorithm for W takes the form [28], [29]:

$$\begin{aligned}\Delta W &= -\eta \overline{\nabla} L(W) \\ &= \eta[y\varphi^T(y)W - \varphi(y)x^T] \end{aligned} \quad (21)$$

where η is a learning rate (small positive constant) and $\varphi(y)$ is non-linear function, noted by

$$\varphi(y) = \frac{1}{a_1}\log(\cosh(a_1 y)) \quad (22)$$

where $1 < a_1 < 2$ is some suitable constant.

In the learning process, the increment ΔW should satisfy the constraint

$$\Delta W W^T + W \Delta W^T = 0 \quad (23)$$

4.2 Feature Extraction

Image representations are often based on discrete linear transformations of the observed data. Consider a black-and-white image whose gray-scale value at the pixel indexed by x and y, denoted by $I(x, y)$. Many basic models in image processing express the image $I(x, y)$ as a linear superposition of some features or basis functions $a_i(x, y)$, that is

$$I(x, y) = \sum_{i=1}^{M} a_i(x, y)s_i \quad (24)$$

where s_i are feature coefficients. These basis functions, $a_i(x, y)$, are able to capture the inherent structure of the iris texture. This, particularity allows us to apply ICA and thus create a set of compact features for an effective recognition task. Alternatively, we can just collect all the pixel values in a single vector X, in which case we can express the representation as in (8) for ICA model. We assume here that the number of transformed components is equal to the number of observed variables. This type of a linear superposition model gives a useful description on a low level support where we can ignore such higher-level nonlinear phenomena such as occlusions. For the sake of simplicity, let us restrict ourselves here to the simple case where the variables

$a_i(x,y)$ form an invertible linear system, that is, the matrix A is square. Then we can invert the system as

$$s_i = \sum_{x,y} w_i(x,y) I(x,y) \qquad (25)$$

where the w_i denote the inverse filters of ICA.

In practice, we cannot model a whole image using the model in (24). Rather, we apply it on image patches or windows [23]. Thus we partition the image into patches of $n \times n$ pixels and model the patches with the model in (24). However, care must then be taken to avoid border effects.

Before extracting the iris features, we note that the ICA application is greatly simplified if the vector X of all iris images is first whitened or sphered. There are two common pre-processing steps. The first step is to center the images as, $X = X - E\{X\}$ in order to make their local mean equal 0. The next step is to apply a whitening transform B to the data such that

$$B = D^{-\frac{1}{2}} E^T \qquad (26)$$

with E corresponds to the eigenvectors of the covariance matrix of X and the diagonal matrix D contains the related eigenvalues. The whitening process helps to uncorrelate the data so that Principal Component Analysis (PCA) can work with a unit variance. The whitened data are used as the input for the Flexible-ICA algorithm [24], demonstrated above, which computes a set of basis vector, w_i from a set of iris images, and the images are projected into the compressed subspace to obtain a set of coefficients, s_i. New test images are then matched to these known coefficients by projecting them onto the basis vectors and finding the closest coefficients in the subspace.

5 Matching

It is very important to present the obtained feature vector in a binary code because it is easier to determine the difference between two binary code-words than between two number vectors. In fact, Boolean vectors are always easier to compare and to manipulate. We have applied a Hamming Distance matching algorithm for the recognition of two samples. It is basically an exclusive OR (XOR) function between two bit patterns. Hamming Distance is a measure, which delineates the differences of iris codes. Every bit of a presented iris code is compared to the every bit of referenced iris code, if the two bits are the same, e.g. two 1's or two 0's, the system assigns a value '0' to that comparison and if the two bits are different, the system assigns a value '1' to that comparison. The formula for iris matching is shown as follows

$$HD = \frac{1}{N} \sum P_i \oplus Q_i \qquad (27)$$

where N is the dimension of feature vector, P_i is the ith component of the presented feature vector, while Q_i is the ith component of the referenced feature vector.

6 Experimental Results and Analysis

This section is concerned with description of the database used in our analysis, the iris recognition process based on ICA, the results obtained and their analysis using some mathematical criteria including a comparative study against some existing methods.

6.1 Iris Database

CASIA-IrisV3 includes three subsets which are labelled as CASIA-IrisV3-Interval, CASIA-IrisV3-Lamp, and CASIA-IrisV3-Twins [21].

CASIA-IrisV3 contains a total of 22,051 iris images taken from more than 700 subjects and 1,500 eyes. All iris images are 8 bit gray-level JPEG files, collected under near infrared illumination. Almost all subjects are Chinese except a few in CASIA-IrisV3-Interval. Because the three data sets were collected at different times, only CASIA-IrisV3-Interval and CASIA-IrisV3-Lamp have a small overlap in subjects, which are used to test our algorithm.

CASIA V3-Interval. contains a total of 2,655 iris images from more than 249 subjects and 396 classes. Iris images were captured in two sessions, with one month interval using specialised digital optics developed by the National Laboratory of Pattern Recognition, China. The captured images with a resolution of 320×280 are original unmasked with very good quality and the iris texture details are extremely clear.

CASIA V3-Lamp. contains a total of 16,213 non-ideal iris images from more than 411 subjects and 819 classes, the collection was taken in one session by using OKI's hand-held iris sensor. The captured images with a resolution of 640×480 have nonlinear deformation due to variations of visible illumination.

6.2 Experimental Steps

The experiments were performed in Matlab 7.3 on a T2330 dual-Core 1.60 GHz CPU with 2048 M RAM. The results have been obtained by using 1530 iris images including 306 classes of *CASIA-IrisV3-Interval* subset and 2052 iris images including 228 classes of *CASIA-IrisV3-Lamp* subset of CASIA Iris Image Database V3. Each iris image should be localised by detecting its inner and outer boundary and its eyelids and eyelashes, unwrapped and converted into its polar equivalent; where a number of data points are selected along each radial line and this is defined as the radial resolution and the number of radial lines going around the iris region is defined as the angular resolution. Then a histogram stretching method was used to obtain a well distributed iris images. Figure 3 gives an example of an iris sample of each subset with its pre-processing steps.

So we have obtained a total of 1530 pre-processed image samples of CASIA-IrisV3-Interval and 2052 of CASIA-IrisV3-Lamp of size of 32×240 pixels used for features extraction process, as illustrated in figure 4, which consists of determining the values of s_i and $w_i(x,y)$ for all i and (x,y), of the representation in (25) by giving a sufficient number iris image patches $I(x,y)$.

First, we consider one image for each class, (i.e. 306 or 228 images if we use CASIA V3-Interval or CASIA V3-Lamp, respectively). These images are partitioned to 10,000 image patches of size of $n \times n$ pixels, which were taken at random locations from the pre-processed images, and gathered by normalising each image patch to column vector of size ($n^2 \times 1$), then held into matrix X of size ($n^2 \times 10,000$).

The dimension of X is reduced to $R \times 10,000$, ($R < n^2$). The separating matrix W is calculated using the whitening process and Flexible-ICA algorithm described above. In the other hand, all iris images of a given subset are partitioned to $n \times n$ image patches and each one is normalised to a column vector and held in $\hat{X}$, this later is projected in stiefel manifold $W(R \times n^2)$ in order to obtain feature vectors S. Then, we generate the corresponding iris code for storage and comparison purposes. The encoding method of iris code is to assign values of 0 or 1 to each IC coefficient

$$Q(S_i) = \begin{cases} 1 & if \quad S_i > 0 \\ 0 & if \quad S_i \leq 0 \end{cases} \tag{28}$$

Finally, we use the Hamming distance to compare two iris codes. The hamming distance is the count of bits different in the two iris codes.

6.3 Evaluation Criteria

To evaluate the similarity of the projected iris images, we have used a corresponding matching metric that should give one range of values when comparing projected iris images of the same eye (intra-class comparisons) and another range of values when comparing the projected iris images created from different irises (inter-class comparisons). These two cases should give distinct and separate values, so that a decision can be made with high confidence as to whether two projected iris images are from the same iris, or from two different irises.

The experiments were completed in a verification mode, the receiver operating characteristic (ROC) curve and *equal error rate* (EER) are used to evaluate the performance of the proposed method. The ROC curve is a *false acceptance rate* (FAR) versus *false rejection rate* (FRR) curve, which measures the accuracy of matching process and shows the overall performance of an algorithm. The FAR is the probability of accepting an imposter as an authorized subject and the FRR is the probability of an authorized subject being incorrectly rejected. Points on this curve denote all possible system operating states in different tradeoffs. The ideal FAR versus FRR curve is a horizontally straight line with zero false rejection rate. The EER is the point where the false acceptance rate and the false rejection rate are equal in value. The smaller the EER is, the better the algorithm.

The accuracy of the system, feature vector size and computational complexity are also used to compare our iris proposed algorithm with the algorithms proposed by Daugman [8], Ma et. al. [16] and Tan et al. [15].

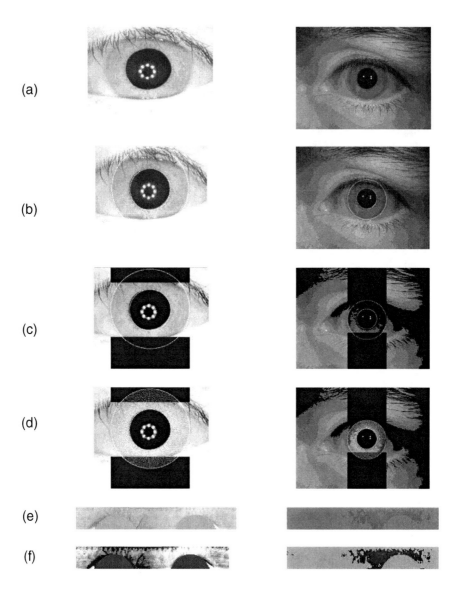

Fig. 3. Iris image pre-processing steps of a sample of each subset of CASIA Iris database, CASIA V3-Interval and CASIA V3-Lamp (*left and right*), (a) original iris, (b) iris localisation, (c) eyelash and eyelids detection, (d) unwrapped iris with a radial resolution of 32 pixels and angular resolution of 240 pixels, (e) normalised and (f) enhanced iris.

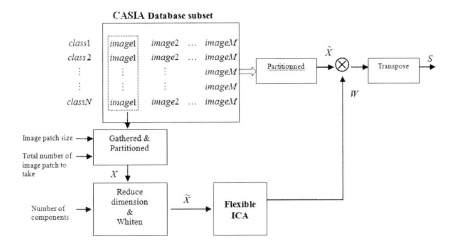

Fig. 4. Block diagram of feature extraction process

6.4 Results and Discussion

To assess the accuracy of the proposed ICA algorithm, each iris image in the database is compared with all the other irises in the CASIA Iris database. For the *CASIA-V3-Interval* subset, the total number of comparisons is 49,725, where the total number of intra-class comparisons is 3,060 and that of inter-class comparisons are 46,665. For the *CASIA-V3-Lamp* subset, the total number of comparisons is 34,086, where the total number of intra-class comparisons is 8,208 and that of inter-class comparisons are 25,878. The intra-class and inter-class distance distribution plot of Hamming Distance calculated by the proposed ICA algorithm of *CASIA-V3-Interval* and *CASIA-V3-Lamp* are showed in figure 5.

Figure 5 shows that the distance between the intra-class and the inter-class distribution is large, indicating a good discriminability of the extracted features. Figure 5 reveals also that the intra-class distance distribution of *CASIA-V3-Lamp* is larger than the intra-class distance distribution of *CASIA-V3-Interval*, the reason being that *CASIA-IrisV3-Lamp* images were taken by the variation of visible illumination with nonlinear deformation which gives bad results of the phase of localisation and normalisation. This is verified by the following verification results.

Table 1 shows EERs for each subset and each reduced size of ICA coefficients, given by $R=\{56, 40, 32, 24, 20, 16, 12, 10, 8\}$ and image patches sizes given by 16×16 and 8×8 pixels of *CASIA V3-Interval* and *CASIA V3-Lamp*.

As mentioned in table 1, EERs obtained by *CASIA-IrisV3-Interval* according to different ICs values, when image patches size is 8×8 pixels, are lower than 0.2%. The reason that *CASIA-IrisV3-Interval* images are very good quality resulting in extremely clear iris texture details. These results are better than results obtained when image patches size is 16×16 pixels. This means that ICs values are very small compared to image patches size and the whitened data fails to capture enough information on the original data, this requires increasing the number ICA coefficients. However, EERs obtained using *CASIA-IrisV3-Lamp* with 16×16 image patches size, which lie between 12.34 % and 15.54 %, are better than those obtained with 8×8 image patches size, which are between 15.69% and 16.89% when the reduced size is higher than 20. An explanation for this behavior can be that the small eigenvalues correspond to high-frequency components and usually encodes noise. We can see that the error rate increases when ICA coefficients decrease, but when the information is strongly affected by noise according to some coefficients, the performance does not always decrease with the reduction of ICA coefficients, and this leads to an unstable ICA estimation.

The performance of ICA is evaluated by taking the best result shown in table 1 which will be compared with algorithms proposed by Daugman [9] [19], Ma et. al. [16] and Tan et al. [15] using the *CASIA-V3-Interval* iris image database [21]. These previous methods for iris recognition mainly focus on feature representation and matching. Therefore, we only analyze and compare the accuracy, efficiency and computational complexity of feature representation of these methods. The method proposed by Daugman [9] is probably the best-known. Daugman used phase information to represent local shape of the iris details. He projected each small local region onto a bank of Gabor filters, and then quantized the resulting phasor denoted by a complex-valued coefficient to one of the four quadrants in the complex plane. In essence, Daugman analyzed the iris texture by computing and quantizing the similarity between the quadrature wavelets and each local region, which requires that the size of the local region must be small enough to achieve high accuracy, the dimensionality of the feature vector is 2,048 components. *Ma et al.* method [16] constructs a set of intensity signals to contain the most important details of the iris and makes use of stable and reliable local variations of the intensity signals as features, their method contains about 660 components, this is because that their method only records the position of local sharp variations as features and contains less redundant information. In [15], *Tan et al.* utilize multichannel spatial filters to extract texture features of the iris within a wider frequency range, this indicates that the extracted features are more discriminating, they extract local characteristics of the iris from the viewpoint of texture analysis, and the dimensionality of the feature vector is 1,600 components. Figure 6 shows the ROC curves of such verification algorithms.

From the results shown in Figure 6, we can see that our proposed method has the best performance, followed by both Ma et al. and Daugman methods which are slightly better than the method of Tan et al... Our proposed method is based on flexible-ICA algorithm which extracts global features in pre-processing step that reduces dimensions for obtaining ICA components for iris; ICA explores independent components of fine iris features. These components of ICA are statistically independent, which reflect iris detail information (such as freckles, coronas, strips, furrows, crypts, and so on) change, whose distribution indicates iris individual difference for each class. So, the local basis images obtained with ICA can lead to more precise representations.

Since ICA reduces significantly the size of iris code, this leads to decrease of processing time. Table 2 shows that our method consumes less time than others, followed by both Tan and Ma methods which are based on 1-D signal analysis. However, Daugman method involves 2-D mathematical operation. These comparisons indicate that our algorithm has an effective and emerging performance in iris recognition.

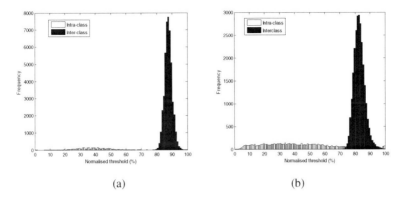

(a)   (b)

Fig. 5. Results of intra-class and inter-class distributions of (*a*) CASIA V3-Interval and (*b*) CASIA V3-Lamp

Table 1. Performance evaluation according to numbers of independent component and image patches sizes of CASIA V3-Interval and CASIA V3-Lamp

Database		CASIA-V3-Interval		CASIA-V3-Lamp	
Win. size		8x8	16x16	8x8	16x16
ICs	56	0.10 %	0.03 %	15.69 %	12.48 %
	40	0.03 %	0.10 %	16.89 %	12.34 %
	32	0.16 %	0.03 %	16.88 %	15.54 %
	24	0.10 %	0.34 %	16.22 %	13.62 %
	20	0.14 %	0.86 %	16.19 %	18.09 %
	16	0.04 %	0.45 %	16.82 %	17.08 %
	12	0.16 %	3.95 %	19.93 %	20.83 %
	10	0.13 %	3.72 %	16.76 %	19.28 %
	8	0.13 %	2.25 %	18.34 %	23.96 %

Table 2. Performance comparison of the algorithms

Methods	Feature vector size (bit/image)	Performance (%)	Computational Complexity (*ms*)
Daugman [9]	2048	0.08	285
Ma et al. [16]	660	0.07	95
Tan et al. [15]	1600	0.48	80.3
Proposed ICA	960	0.04	31.2

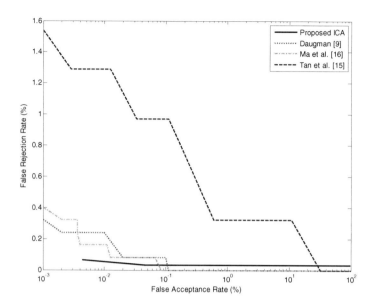

Fig. 6. Comparison of ROC curves

7 Conclusion

In this paper, we have introduced an effective iris recognition system based on Hough transform for iris localisation, Daugman's cartesian to polar transform for normalisation, ICA for feature extraction with partition of iris images into patches, and hamming distance for matching. Two iris image subsets of CASIA iris V3 database have been used to evaluate the performance of our system. Flexible-ICA algorithm, which improves the quality of separation introducing a better density matching and allows a faster learning, has been adopted for computing the ICs.

The results obtained suggest that ICA could perform well for such application with small size of image patches of iris images with very good quality, but it is insensitive to variant illuminations and noise caused by eyelids and eyelashes, and even for blurred iris images, which is a limitation of ICA. For this, we propose to adopt other existing ICA algorithms, such as Noisy-ICA algorithm, for this application. Through comparing with some existing methods, all experimental results have demonstrated that Flexible-ICA algorithm achieves very good performance in both speed and accuracy.

One of the future directions on iris recognition is on the protection of this system from attacks. We would like to propose a multifactor biometric authentication method based on cryptosystem keys containing biometric signatures.

References

1. Wang, Y., Han, J.Q.: Iris Recognition Using Independent Component Analysis. In: Proceedings of the Fourth International Conference on Machine Learning and Cybernetics, Guangzhou, August 18-21 (2005)
2. Ross, A.A., Nandakumar, K., Jain, A.K.: Handbook of Multibiometrics. Springer Science+Business Media, LLC (2006)
3. Jain, A., Bolle, R., Pankanti, S. (eds.): Biometrics: Personal Identification in a Networked Society. Kluwer, Dordrecht (1999)
4. Zhang, D.: Automated Biometrics: Technologies and Systems. Kluwer, Dordrecht (2000)
5. Mansfield, T., Kelly, G., Chandler, D., Kane, J.: Biometric Product Testing Final Report. issue 1.0, Nat'l Physical Laboratory of UK (2001)
6. Obaidat, M.S., Sadoun, B.: Verification of computer users using keystroke dynamics. IEEE Trans. Syst. Man Cybernet. 27, 261–269 (1997)
7. Wayman, J., Jain, A., Maltoni, D., Maio, D.: Biometric Systems, Technology, Design and Performance Evaluation. Springer, London (2005)
8. Flom, L., Safir, A.: Iris recognition system. U.S. Patent 4,641,349 (1987)
9. Daugman, J.: High confidence visual recognition of persons by a test of statistical independence. IEEE PAMI 15(11), 1148–1161 (1993)
10. Wildes, R.P.: Automated iris recognition: An emerging biometric technology. Proceedings of the IEEE 85, 1348–1363 (1997)
11. Boles, W., Boashash, B.: A human identification technique using images of the iris and wavelet transform. IEEE Trans. Signal Proc. 4, 1185–1188 (1998)
12. Lim, S., Lee, K., Byeon, O., Kim, T.: Efficient iris recognition through improvement of feature vector and classifier. ETRI J. 23(2), 61–70 (2001)
13. Tisse, C., Martin, L., Torres, L., Robert, M.: Person identification technique using human iris recognition. In: Proc. of Vision Interface, pp. 294–299 (2002)
14. Zhu, Y., Tan, T., Wang, Y.: Biometric personal identification based on iris pattern. In: ICPR 2000: the 15th International Conference on Pattern Recognition, Barcelona, Spain, pp. 805–808 (2002)
15. Ma, L., Tan, T., Wang, Y., Zhang, D.: Personel identification based on iris texture analysis. IEEE Trans. on Pattern Analysis and Machine Intelligence 25(12), 1519–1533 (2003)
16. Ma, L., Tan, T., Wang, Y., Zhang, D.: Efficient Iris Recognition by Characterizing Key Local Variations. IEEE Transactions on Image Processing 13(6), 739–750 (2004)
17. Du, Y., Ives, R., Chang, C.-I., Etter, D., Welch, T.: Information divergence-based iris pattern recognition for automatic human identification. In: Proc. SPIE, vol. 5404 (2004)
18. Du, Y., Ives, R., Chang, C.-I., Ren, H., D'Amico, F., Jensen, J.: A new hyperspectral discrimination measure for spectral similarity. Optical Engineering 43(8) (2004)
19. Daugman, J.G.: How iris recognition works. IEEE Trans. Circuits and Syst. for Video Tech. 14(1), 21–30 (2004)
20. Stone, J.V.: Independent Component Analysis. A Tutorial Introduction. A Bradford Book, MIT Press (2004)
21. Download the application form at the website, http://www.cbsr.ia.ac.cn/IrisDatabase.htm
22. Sanderson, S., Erbetta, J.: Authentication for secure environments based on iris scanning technology. IEEE Colloquium on Visual Biometrics (2000)
23. Hyvarinen, A., Karhunen, J., Oja, E.: Independent Component Analysis. John Wiley, Chichester (2001)

24. Choï, S., Cichocki, A., Amari, S.: Adaptative Blind Signal and Image Processing: Learning Algorithms Applications. John Wiley & Sons, Chichester (2002)
25. Cover, T.M., Thomas, J.A.: Elements of Information Theory. Wiley, Chichester (1991)
26. Papoulis, A.: Probability, Random Variables, and Stochastic Processes, 3rd edn. McGraw-Hill, New York (1991)
27. Comon, P.: Independent component analysis—a new concept? Signal Processing 36, 287–314 (1994)
28. Amari, S.: Natural Gradient for over under complete bases in ICA. Neural Computation 11(8), 1875–1883 (1999)
29. Vigliano, D., Parisi, R., Uncini, A.: A flexible ICA approach to a novel BSS convolutive nonlinear problem: preliminary results. Biological and Artificial Intelligence Environments Part 3, 217–224 (2005), doi:10.1007/1-4020-3432-6_26
30. Meena, B.R.: Personal Identification based on Iris Patterns. Ph.D Thesis, Department of Computer Science and Engineering, IndianInstitute of Technology, Kanpur (2004)
31. Vatsa, M., Singh, R., Noore, A.: Reducing the False Rejection Rate of Iris Recognition Using Textural and Topological Features. International Journal of Signal Processing 2(2), 66–72 (2005)
32. Monro, D.M., Rakshit, S., Zhang, D.: DCT-Based Iris Recognition. IEEE Transactions on Pattern analysis and Machine Intelligence 29(4), 586–595 (2007)
33. Dorairaj, V., Fahmy, G., Schmid, N.: Performance evaluation of iris based recognition system implementing PCA and ICA techniques. In: Proc. of SPIE Conference on Biometric Technology for Human Identification (2005)
34. Bae, K., Noh, S., Kim, J.: Iris feature extraction using independent component analysis. In: Kittler, J., Nixon, M.S. (eds.) AVBPA 2003. LNCS, vol. 2688, pp. 838–844. Springer, Heidelberg (2003)
35. Oyster, C.: The Human Eye Structure and Function. Sinauer Associates (1999)
36. US NLM/NIH Medline Plus. Cataract, http://www.nlm.nih.gov/medlineplus/ency/article/001001.htm (accessed October 2006)
37. US NLM/NIH Medline Plus. Glaucoma, http://www.nlm.nih.gov/medlineplus/ency/article/001620.htm (accessed October 2006)
38. European Commission. Biometrics at the frontiers: Assessing the impact on society. Institute for Prospective Technological Studies, Technical Report EUR 21585 EN (European Commission Director-General Joint Research Centre) (February 2005)
39. US NLM/NIH Medline Plus. Albinism, http://www.nlm.nih.gov/medlineplus/ency/article/001479.htm (accessed January 2007)
40. Hyvärinen, A.: Fast and robust fixed-point algorithms for independent component analysis. IEEE Transactions on Neural Networks 10(3), 626–634 (1999)

Author Index

Alimazighi, Zaia 95
Amarouche, Idir Amine 95

Barhamgi, Mahmoud 95
Baumgartner, Christian 148
Bechini, Alessio 15
Benslimane, Djamal 95
Bonfante, Paola 124
Bouraoui, Imen 188
Bouridane, Ahmed 188

Chitroub, Salim 188
Cordero, Francesca 124
Couto, Francisco 40

Dander, Andreas 148

Ghignone, Stefano 124
Giannini, Riccardo 15

Handler, Michael 148
Herz, Andreas 1

Ienco, Dino 124

Jamil, Hasan 158

Lanfranco, Luisa 124
Leonardi, Giorgio 124
Li, Chen 73

Meo, Rosa 124
Montani, Stefania 124
Mrissa, Michael 95

Netzer, Michael 148

Pfeifer, Bernhard 148

Ramampiaro, Heri 73
Rautenberg, Philipp L. 1
Roversi, Luca 124

Seger, Michael 148
Silva, Fabrício A.B. 40
Silva, Mário J. 40
Sobolev, Andrey 1

Visconti, Alessia 124

Wachtler, Thomas 1

Zamite, João 40